EIT
Chemical
Review

Second Edition

Dilip K. Das, P. E.
Dr. Rajaram K. Prabhudesai, P. E.

Engineering Press   **Austin, Texas**

ISBN 1-57645-023-6

Printing 5 4 3 2 1

Engineering Press
P.O. Box 200129
Austin, TX 78720-0129

(800)800-1651
FAX (800)700-1651

Contents

After a 30-year career in the environmental engineering area, I served 10 years as the Director of Examinations for the National Council of Examiners for Engineering and Surveying (NCEES). The NCEES develops and grades the examinations that all 50 states and 5 jurisdictions of the USA use in licensing engineers and land surveyors. During my time at the NCEES, approximately 500,000 candidates sat for the Fundamentals (FE/EIT) and about 300,000 took the Principles and Practice (PE) examinations.

Engineering Press has provided top quality products over a period of years with guidance to candidates who are preparing to take the exams. Their publications are kept up-to-date with the many changes that have been made in the exams without overloading them with extraneous information.

The specifications for all the exams are developed by surveys of the practicing professional engineers and the education professionals who teach in the accredited engineering programs at universities and colleges throughout the United States. Item (problem, question) writers from all over the country participate in developing the items that are used on all exams. They must be licensed in at least one of the 55 states or jurisdictions and subject matter experts in a particular area. They come from all geographical areas and represent different areas of practice such as consulting, private practice, education and government. Many of them serve on the various state licensing boards.

The actual exams are assembled by committee members with the same representation as above. Committee members are also expected to write items. It is estimated that each item on a particular test will have been reviewed by at least 5 - 10 individuals over a one to one and a half year period of time before it appears on an examination. All reviews are conducted to ensure the items have only one correct answer (multiple-choice), are clear, unambiguous and of the appropriate difficulty level. In spite of this careful process, flawed items are occasionally found on the tests.

During scoring process and before any results are sent to the state boards, various statistical methods are used to provide a quality check on the performance of each item on the test. Sometimes seriously flawed items may be eliminated from scoring or other corrections made. In other words, credit may be given to all candidates regardless of their response. All decisions are made by subject matter experts with the best interest of the candidates, consistent with public protection, in mind.

I encourage you to take advantage of the publications provided by Engineering Press for success on the first attempt. If you have specific questions, use the web site at: http://www.engrpress.com. For detailed information on the exams refer to the NCEES web site at: http://www.ncees.com.

J. Earl Herndon Jr., P. E., DEE
Former Director of Exams

Preface

The Fundamentals of Engineering/Engineer-In-Training (FE/EIT) exam has dramatically changed. For about the last fifty years the 8-hour exam has been a test of a dozen topics that are the core of every engineer's education: Math-Chem-Physics-Statics-Dynamics-Fluids-Thermo-Circuits-Engr Econ-Mat Sci-Computers and Mech of Materials. All that changed in the Fall of 1996. Now the exam has two distinct parts to it. The morning four-hour portion is pretty much like the old exam with about the same dozen topics. This book has been revised in this edition to closely follow the afternoon topics and the afternoon exam content of the *Fundamentals of Engineering Reference Handbook*.

The afternoon four-hour portion of the exam is new and different. In fact there are six different afternoon exams. Each one covers a different branch of engineering: civil, mechanical, electrical, chemical, industrial and general. While the morning exam covers the first 2-2½ years of an engineer's education, the afternoon exam covers the last 1½-2 years of an engineers education in his specific branch of engineering. This means that an orderly review in preparation for the exam has two components and requires two books. Everyone can use the *Engineer-In-Training License Review* book for the morning exam, as can general engineers for the afternoon exam. All other engineers will need one of the following:

EIT Civil Review	*EIT Chemical Review*
EIT Mechanical Review	*EIT Industrial Review*
EIT Electrical Review	

To prepare these books I assembled a carefully selected team of professionals, with each professional preparing that portion of the book that relates to his expertise. I am most grateful to Dr. Prabhudesai and Dilip Das for devoting a lot of time and effort to writing this book. If you encounter any errors in this book, we would appreciate being notified. Notes and comments may be sent directly to me, c/o Engineering Press, P.O. Box 200129, Austin, TX 78720-0129.

These books, in their previous editions, have helped tens of thousands of engineers prepare for and pass the FE/EIT exam. The team of engineering professors and I have worked diligently to provide you the best possible review to quickly and efficiently prepare for this changed exam. We wish you good luck in the exam.

Donald G. Newnan

INTRODUCTION

BECOMING A PROFESSIONAL ENGINEER

To achieve registration as a Professional Engineer, there are four distinct steps: 1) education, 2) the Fundamentals of Engineering/Engineer-In-Training (FE/EIT) exam, 3) professional experience, and 4) the professional engineer exam. These steps are described in the following sections.

Education

Generally, no college degree is required to be eligible to take the FE/EIT exam. The exact rules vary, but all states allow engineering students to take the FE/EIT exam before they graduate, usually in their senior year. Some states, in fact, have no education requirement at all. One merely need apply and pay the application fee. Perhaps the best time to take the exam is immediately following completion of related course work. For most engineering students, this will be in the senior year.

Fundamentals of Engineering/
Engineer-In-Training Examination

This eight-hour, multiple-choice examination is known by a variety of names: Fundamentals of Engineering, Engineer-In-Training (EIT.), or Intern Engineer, but no matter what it is called, the exam is the same in all states. It is prepared and graded by the National Council of Examiners for Engineering and Surveying (NCEES).

Experience

States that allow engineering seniors to take the FE/EIT exam have no experience requirement. These same states, however, generally will allow other applicants to substitute acceptable experience for course work. Still other states may allow one to take the FE/EIT exam without any education or experience requirements.

Professional Engineer Examination

The second national exam is called Principles and Practice of Engineering by NCEES, but many refer to it as the Professional Engineer exam or P.E. exam. All states, plus Guam, the District of Columbia, and Puerto Rico use the same NCEES exam. Typically, four years of acceptable experience is required before one can take the Professional Engineer exam, but the requirement may vary from state to state. Review materials for this exam are found in other engineering license review books.

FUNDAMENTALS OF ENGINEERING/
ENGINEER-IN-TRAINING EXAMINATION

Background

Laws have been passed that regulate the practice of engineering to protect the public from incompetent practitioners. Beginning about 1907 the individual states began passing *title* acts regulating who could call themselves an engineer. As the laws were strengthened, the *practice* of engineering was limited to those who were registered engineers, or to those working under the supervision of a registered engineer. Originally the laws were limited to civil engineering, but over time they have evolved so that the *titles* and sometimes the *practice* of most branches of engineering are included. There is no national registration law; registration is based on individual state laws and is administered by boards of registration in each state.

Examination Development

Initially, the states wrote their own examinations, but beginning in 1966 the NCEES took over the task for some of the states. Presently the NCEES exams are used by all states. Now it is easy for engineers that move from one state to another to achieve registration in the new state. About 50,000 engineers take the FE/EIT exam annually. This equals about half the engineers graduated in the U.S. each year.

The development of the FE/EIT exam is the responsibility of the NCEES Committee on Examinations for Professional Engineers. The committee is composed of people from industry, consulting, and education, some of whom are subject-matter experts. The test is intended to evaluate an individual's understanding of basic sciences as well as engineering sciences along with one's ability to apply this knowledge to engineering problems. Every five years or so, NCEES conducts an engineering task analysis survey. People in industry, consulting, and education are surveyed to review engineering practice and the knowledge needed for it. From this, NCEES develops what they call a "matrix of knowledge" that forms the basis for the FE/EIT exam structure described in the next section.

The actual exam questions are prepared by the NCEES committee members, subject matter experts, and other volunteers. All people participating must hold professional registration. When the questions have been written, they are circulated for review in workshop meetings and by mail. Both single- and multi-part questions are used; the latter are arranged so the solution of each succeeding part does not depend on the correct solution of a previous part. All problems are four-way multiple choice.

Examination Structure

The FE/EIT exam is divided into a morning four-hour section and an afternoon four-hour section. The Fundamentals of Engineering Examination in the afternoon period has changed to a discipline specific examination. Six exams are prepared in the afternoon test booklet, one for each of the following five disciplines: Civil, Mechanical, Electrical, Chemical, Industrial; and a General topic for those examinees not covered by the five engineering disciplines. The test consists of 60 problems in the topic. This may be of great benefit to those specializing in one of the five major engineering disciplines. Most of the test's topics cover the third and fourth year of your college courses. These are the courses which you will use for the balance of your engineering career, so the test becomes focused to your own needs. Graduate engineers will find the PM discipline test to their advantage as the broad fundamentals test usually causes many engineers to do a good deal of review of the earliest class-work.

The major subjects for the afternoon sections are:

Civil	No.Probs.
Soil Mechanics & foundations	6
Structural Analysis, frames, trusses	6
Hydraulics & hydro systems	6
Structural Design, concrete, steel	6
Environmental Engineering, waste water, solid waste treatment	6
Transportation facilities, highways railways, airports	6
Water purification and treatment	6
Surveying	6
Computer and Numerical methods	6
Legal & professional aspects	3
Construction Management	3

Mechanical	
Mechanical Design	6
Dynamic Systems,vibration,kinematics	6
Thermodynamics	6
Heat Transfer	6
Fluid Mechanics	6
Stress Analysis	6
Measurement and Instrumentation	6
Material Behavior/ Processing	3
Computer, Automation, Robotics and Numerical methods	3
Energy Conversion and Power Plants	3
Automatic control	3
Refrigeration and HVAC	3
Fans, Pumps, and Compressors	3

Electrical	
Digital Systems	6
Analog Electronic Circuits	6
Electromagnetic Theory & Applications	6
Network Analysis	6
Control Systems Theory and Analysis	6
Solid State Electronics and Devices	6
Communications Theory	6
Signal Processing	3
Power Systems	3
Computer Hardware Engineering	3
Computer Software Engineering	3
Instrumentation	3
Computer and Numerical Methods	3

Chemical	No.Probs..
Material & Energy Balances	9
Chemical Thermodynamics	6
Mass Transfer	6
Chemical Reaction Engineering	6
Process Design and Econ Evaluation	6
Heat Transfer	6
Transport Phenomenon	6
Process Control	3
Process Equipment Design	3
Computer and Numerical Methods	3
Process Safety	3
Pollution Prevention	3

Industrial	
Production Planning & Scheduling	3
Engineering Economics	3
Engineering Statistics	3
Statistical Quality Control	3
Manufacturing Processes	3
Mathematical Optimization-Modeling	3
Simulation	3
Facility design and location	3
Work Performance and Methods	3
Manufacturing Systems Design	3
Industrial Ergonomics	3
Industrial Cost Analysis	3
Material Handling System Design	3
Total Quality Management	3
Computer Computations-Modeling	3
Queuing Theory and Modeling	3
Design of Industrial Experiments	3
Industrial Management	3
Information System Design	3
Productivity Measurement	3

General	
Mathematics	12
Electrical Circuits	6
Statics	6
Thermodynamics	6
Chemistry	4
Dynamics	4
Mechanics of Materials	4
Material Science	3
Computers	3
Ethics	3
Engineering Econ	3

Taking the Examination

Examination Dates

The National Council of Examiners for Engineering and Surveying (NCEES) prepares FE/EIT exams for use on a Saturday in April and October each year. Some state boards administer the exam twice a year; others offer the exam only once a year. The scheduled exam dates are:

	April	October
1998	25	31
1999	24	30
2000	15	28

Those wishing to take the exam must apply to their state board several months before the exam date. We recommend that Civil, Mechanical, Electrical, Chemical, and Industrial engineers take the discipline specific exam. All others should take the General examination.

Examination Procedure

Before the morning four-hour session begins, the proctors will pass out exam booklets and a scoring sheet to each examinee. An HB or #2 pencil is to be used to record answers. Space is provided on each page of the examination booklet for scratch work. The scratch work will *not* be considered in the scoring.

The examination is closed book. You may not bring any reference materials with you to the exam. To replace your own materials, NCEES has prepared a *Fundamentals of Engineering (FE) Reference Handbook.* The handbook contains engineering, scientific and mathematical formulas and tables for use in the examination. Examinees will receive the handbook from their state registration board prior to the examination. The *FE Reference Handbook* also is included in the exam materials distributed at the beginning of each four-hour exam period. If you do not already have a copy of the *FE Reference Handbook* you should purchase one right away.

There are three versions (A, B, and C) of the exam. These have the major subjects presented in a different order to reduce the possibility of examinees copying from one another. The first subject on your exam, for example, might be fluid mechanics, while the exam of the person next to you may have electrical circuits as the first subject.

The afternoon session begins following a one-hour lunch break. The afternoon exam booklets will be distributed along with a scoring sheet. At the beginning of the afternoon test period, the examinee will mark the answer sheet as to which exam he is taking. There will be (60 multiple choice questions, each of which carries twice the grading weight of the morning exam questions.

If you answer all questions more than 30 minutes early, you may turn in the exam materials and leave. In the last 30 minutes, however, you must remain to the end of the exam period to ensure a quiet environment for all those still working, and to ensure an orderly collection of materials.

Examination-Taking Suggestions

Those familiar with the psychology of examinations have several suggestions for examinees:

1. There are really two skills that examinees can develop and sharpen. One is the skill of illustrating one's knowledge. The other is the skill of familiarization with examination structure and procedure. The first can be enhanced by a systematic review of the subject matter. The second, exam-taking skills, can be improved by practice with sample problems-that is, problems that are presented in the exam format with similar content and level of difficulty.

2. Examinees should answer every problem, even if it is necessary to guess. There is no penalty for guessing. The best approach to guessing is to first eliminate the one or two obviously incorrect answers among the four alternatives. If this can be done, the chance of selecting a correct answer obviously improves from 1 in 4 to 1 in 2 or 3.

3. Plan ahead with a strategy and a time allocation. There are 120 morning problems in 12 major subject areas. Compute how much time you will allow for each of the 12 subject areas. You might allocate a little less time per problem for those areas in which you are most proficient, leaving a little more time in subjects that are more difficult for you. Your time plan should include a reserve block for especially difficult problems, for checking your scoring sheet, and finally to make last-minute guesses on problems you did not work. A time plan gives you the confidence of being in control. Misallocation of time for the exam can be a serious mistake. Your strategy might also include time allotments for two passes through the exam ; the first to work all problems for which answers are obvious to you; the second to return to the more complex, time-consuming problems and the ones at which you might need to guess.

4. Read all four multiple choice answers before making a selection. The first answer in a multiple choice question is sometimes a plausible decoy-not the best answer.

5. Do not change an answer unless you are absolutely certain you have made a mistake. Your first reaction is likely to be correct.

6. If time permits, check your work.

7. Do not sit next to a friend, a window, or other potential distraction.

License Review Books

There are two rules that we suggest you follow in selecting license review books to ensure that you obtain up-to-date materials:

1. Consider only the purchase of materials that have a 1993 or later copyright. Prior to October, 1993 reference books could be brought to the exam, and some review books added unnecessary reference materials to get them accepted as reference books. That strategy was successful then, but it makes no sense now. Nothing can be taken into the exam, and the unnecessary material in these older books diverts attention from the exam content. Look for books that cover what you need to know--no more, and no less.

2. Even if a license review book has a recent copyright date, is the content up to date? Books that ignore the multiple-choice format, or include irrelevant material, are not up to date.

Textbooks

If you still have your university textbooks, they can be useful in preparing for the exam, unless they are too out of date. To a great extent the books will be like old friends with familiar notation. You probably need both textbooks and license review books for efficient study and review. You must, however, pay close attention to the *FE Reference Handbook and* the notation used in it, because it is the only book you will have in the exam.

Examination Day Preparations

There is no doubt that the exam day will be a stressful and tiring one. You should take steps to eliminate the possibility of unpleasant surprises. If at all possible, visit the examination site ahead of time. Try to determine such items as:

1. How much time should I allow for travel to the exam on that day. Plan to arrive about 15 minutes early. That way you will have ample time, but not too much time. Arriving too early and mingling with others who also are anxious, can increase your anxiety and nervousness.

2. Where will I park?

3. How does the exam site look. Will I have ample work space? Will it be overly bright (sunglasses) or cold (sweater), or noisy (earplugs)? Would a cushion make the chair more comfortable?

4. Where is the drinking fountain? Lavatory facilities? Pay phone?

5. What about food? Should I take something, along for energy in the exam? A light bag lunch during the break probably makes sense.

Items to Take to the Examination

● Calculator-NCEES says you may bring a battery-operated, silent, non-printing calculator. You need to determine whether or not your state permits preprogrammed calculators. Bring extra batteries for your calculator just in case; and many people feel that bringing a second calculator is also a very good idea.

● Clock-You must have a time plan and a clock or wristwatch.

● Pencils-You should consider using mechanical pencils so you don't have to run around sharpening pencils. And you surely do not want to drag along a pencil sharpener.

● Eraser-Try a couple to decide what to bring along. You must be able to change answers on the multiple choice answer sheet, and that means a good eraser.

● Exam Assignment Paperwork-Take along the letter assigning you to the exam at the specified location to prove that you are the registered person. Also bring something with your name and picture (driver's license or identification card).

● Items Suggested By Your Advance Visit--If you visit the exam site, it probably will suggest an item or two that you need to add to your list.

● Clothes-Plan to wear comfortable clothes. You probably will do better if you are slightly cool, so it is wise to wear layered clothing.

Examination Scoring
The questions are machine-scored by scanning. The answer sheets are checked for errors by computer. Marking two answers to a question, for example, will be detected and no credit given.

Passing the FE/EIT Examination
The 120 morning and 60 double-weighted afternoon problems yield a raw score of 240. The passing score is set as the equivalent of a raw score of 124 out of 280 on the October, 1990 examination. To translate this established level of difficulty to new examinations, a portion of each exam contains problems given on previous exams. An analysis of this subset of problems can be used to scale the new exam raw score to the same level of difficulty as 124 on the October, 1990 exam. This produces a standardized exam where the percentage of applicants passing, may vary from exam to exam, but they all demonstrate an established level of competency. Since most state licensing laws require 70 on the exam to pass, the raw passing score (equivalent to 124 on the October, 1990 exam) is converted to 70 and reported to the states accordingly. Higher raw scores are reported proportionally higher than 70; lower raw scores as proportionally lower than 70. What does it all mean? You are faced with 120 single-weighted morning problems and 60 double-weighted afternoon problems which makes a maximum possible score of 240. You must correctly answer problems with an equivalent score of about 124 to receive a passing score. So, you do *not* need to answer 70% of the problems correctly; you will get a converted score of 70 if you answer about 45% of the problems correctly.

The converted scores are reported to the individual state boards in about two months, along with the recommended pass or fail status of each applicant. The state board is the final authority of whether an applicant has passed or failed the exam.

Although there is some variation from exam to exam, the following gives the approximate passing rates:

Applicant's Degree	Percent Passing Exam
Engineering from accredited school	62%
Engineering from non-accredited school	50
Engineering Technology from accredited school	42
Engineering Technology from non-accredited school	33
Non-Graduates	36
First-time Applicants	67
All Applicants	56

If you do not pass the exam on your first attempt, you can re-apply and take it again.

THIS BOOK

This book has been organized to cover the afternoon portion of the Fundamentals of Engineering/Engineer-In-Training (FE/EIT) examination, with nothing added and hopefully nothing left out. Each of the topic authors has created a chapter that reflects the essential material at the level of the FE/ EIT exam.

EIT
Chemical
Review

CHAPTER 1
DIMENSIONS and UNITS

Dimensions are names used to describe the characteristics of a physical quantity. Examples of dimensions are length, mass, time, temperature, and electric charge. Units are the basic standards of measuring the magnitudes of physical quantities. For example, meter is a unit for the measurement of length, and second is a unit for measuring time. Units are two types : fundamental and derived. Fundamental units are independently defined whereas the derived units are expressed in terms of the fundamental units. Various unit systems differ from one another in the choice of fundamental dimensions and units (Table 1-1).

Table 1-1: Dimension and Units

Dimension:	Symbol	Metric or SI system: Unit	British Unit	USCS Unit
Length	L	Meter	foot	foot
Mass	M	kg	slug*	lb_m
Time	t	Second	Second	Second
Temperature	T	degree K	$^0 F$	$^0 F$
Force	F *	Newton (N)*	poundal	lb_f

*denotes a derived unit in the system

Units of Force:

In the SI system, $1 \text{ Newton} = (1 \text{ kg}) \times (1 \text{ m} / \text{s}^2)$

In British Engineering system, $1 \text{ slug} = \dfrac{1 \ pound\text{-}force}{1 \ ft/s^2}$

In fps system, $1 \text{ poundal} = (1 \text{ lb-mass}) \times (1 \text{ ft/s}^2)$

Comparison of British and fps systems shows:

$1 \text{ slug} = 32.174 \ lb_m$

$1 \text{ lb-force (lb}_f) = 32.174 \text{ poundals}$

The US customary system ,

$1 \text{ lb-force (lb}_f) = \dfrac{(1 \ lb_m)(1 \ ft/s^2)}{g_c}$

where Newton's law proportionality factor $g_c = 32.174 \ lb_m.ft/s^2.lb_f$

Molar Units:
A mole of a substance = molecular mass expressed in given units. Molecular mass is also commonly called molecular weight although weight implies force.

1 kilogram-mole = molecular weight in kilograms,
1 pound-mole = molecular weight in pounds.
1 g-mole = molecular weight expressed in grams,
1 kmol = 1000 g-mol

For practical purposes, multiples or submultiples of the same unit are used and are given special names. For example , 1 ft = 12 inches. Here an inch is a unit which has a value of length equal to 1/12 th of one foot. The metric multiplier factors are given in Table 1-2 and some basic conversion factors are given in Table 1-3.

Table 1-2: Multipliers of Standards Units SI or Metric System

Multiplier	prefix of standard unit	symbol	multiplier	prefix of standard unit	symbol
10^9	giga	G	10^{-9}	nano	n
10^6	mega	M	10^{-6}	micro	μ
10^3	kilo	k	10^{-3}	milli	m
10^2	hecto	h	10^{-2}	centi	c
10^1	deka	da	10^{-1}	deci	d

Table 1-3: Some basic conversion factors

Length: 1 meter = 3.281 ft   1 inch = 2.54 cm. = 25.4 mm
1 mile = 5280 ft = 1.609 km   1 foot = 30.48 cm
1 foot = 12 inches

Mass: 1 kg = 2.2046 lb_m 1 lb_m = 453.6 gm = 0.4536 kg
1 U.S. ton = 2000 lb_m 1 ton (British) = 2240 lb_m
1 metric ton = 1000 kg   1 slug = 32.174 lb_m

Time: 1 second = (1/3600) h = (1/60) minute

Force: 1 lb_f = 32.174 poundals. = 4.448 N
1 Newton = 10^5 dyne   1 poundal = 0.138 newton

Temperature :   degree Kelvin = ^0C + 273.16 , ^0R = ^0F + 459.7

Pressure :   1 bar = 1×10^5 Pa = 1 N/m^2 ,   1 atm. = 1.013 Bar = 1.03 kg/cm^2 ,
1 atm. = 760 mm Hg .

Thermal units: 1 calorie = 4.1868 J   1 kcal = 4186.8 J
1 Btu = 0.252 kcal = 778.2 $ft.lb_f$

In practice, one uses conversion factors already tabulated in various references[1,2,3,4] . some factors that are required very often are given in Table 1-4.

PRACTICE PROBLEMS :

1-1: Choose the correct answer for each of the following questions.

(a) In addition to length, time and pound-force, mass is also a fundamental dimension in British
engineering system.   (1) True   (2) False

(b) Dimensional consistency of an equation guarantees its accuracy.   (1) True   (2) False

(c) Conversion factors when utilized are dimensionless.   (1) True   (2) False

(d) Candela is the fundamental unit of luminous intensity in the SI system.   (1) True   (2) False.
(e) The following equation relating the area of a right circular cone of altitude h and base radius of r is
dimensionally consistent. $a = \pi r^2 + \pi r (r^2 + h^2)^{1/2}$ (1) True   (2) False

Table 1-4 : Conversion Factors

Multiply	By	To obtain	Multiply	By	To obtain
Atm std.	14.7	psi	ft-lb$_f$	0.32	cal
Atm std	760	mm Hg	ft-lb$_f$	1.36	joule(J)
Atm std	29.92	in Hg	ft-lb$_f$/s	1.818×10^{-3}	hp
Atm std	33.9	ft of water	gal	3.785	liters(L)
Atm std	1.013×10^5	Pa	hp	33,000	ft-lb$_f$ /min
Atm std	1.013	Bar	hp	42.4	Btu/min
Atm std	1.03	kg/cm^2	hp	745.7	watt(W)
Atm std	101.3	kPa	hp-h	2,545	Btu
Bar	1×10^5	Pa	hp-h	2.68×10^6	joule(J)
Bar	1.0197	kg/cm^2	joule(J)	9.478×10^{-4}	Btu
Bar	14.507	psi	joule(J)	0.74	ft-lb$_f$
Btu	1,054.8	Joule(j)	joule(J)	1	N-m
Btu	2.928×10^{-4}	kWh	joule(J)/s	1	watt(W)
Btu	778.2	ft.lb$_f$	J/(s-m^2-^{0}C)	1	W/m^2-^{0}C
Btu	0.252	kcal	kcal	3.9685	Btu
Btu/h	3.93×10^4	hp	kcal	1.56×10^{-3}	hp-h
Btu/h	0.293	watt(W)	kcal	4.186	joule(J)
Btu/h	0.22	ft.lb$_f$ /s	kN/m^2	0.295	in Hg
Btu/h-ft$^{2 \cdot 0}$ F	4.88	kcal/h-m^2-^{0}C	kPa	0.15	psi
Btu/h-ft^2 - ^{0} F	5.67×10^{-3}	kW/m^2-^{0}C	kW	1.34	hp
Btu/h-ft^2 - ^{0} F	5.68	J/(s-m^2-^{0}C)	kW	737.6	ft.lb$_f$/s
Btu/(h-ft$^{2\,0}$ F/ft)	1.49	kcal/h-^{0}C -m	kWh	3,413	Btu
Btu/(h-ft$^{2\,0}$ F/ft)	1.73	J/s-m-^{0}C	kWh	3.6×10^6	joule(J)
Btu/lb - ^{0} F	1	kcal / kg- ^{0}C	kip(K)	1,000	lb$_f$
cP	1×10^{-3}	N-s/m^2	mile	1.609	km
cP	2.42	lb/h-ft	micron	1×10^{-6}	meter
cP	0.000672	lb/s-ft	poundal	0.14	Newton
cP	3.6	kg/h-m	watts	3.41	Btu/h
cP	1	mPa-s	watts	1	Joule/s
ft^3	7.48	gal	W/m^2	0.317	Btu/h-ft^2
ft-lb$_f$	1.285×10^{-3}	Btu	W/m^2-K	0.1761	Btu/h-ft^2 - ^{0} F
ft-lb$_f$	3.766×10^{-7}	kWh	W/(m^2-K-/m)	0.58	Btu/(h-ft^2- ^{0} F/ft)

(f) Specific weight of a substance is defined as its weight per unit volume. When SI system of units is used, the specific weight w of a substance is given by $w = \rho g$.   (1) True   (2) False

(g) A stone having a mass of 10 kg, is moving with a velocity of 3 m/s . Its kinetic energy is therefore 0.045 kJ.   (1) True   (2) False

(h) The pressure gauge on a pipe line in a plant reads 4 bar. The barometer reads 752.3 mm Hg. The pipe line pressure is therefore 5. bara.   (1) True   (2) False

(i) Water is fed to a steam boiler at a rate of 455 m³/h. Water analysis shows its oxygen content to be 8 ppm. The total dissolved oxygen fed with the water is therefore 3.64 kg/h.   (1) True   (2) False

(j) Heat input into a vessel due to external fire is given by $Q = 21000\ (A)^{0.82}$ wherein Q is in Btu/h and A is wetted area of the vessel in ft². The heat input in kW is therefore given by

$$Q_m = 43.17\ (A_m)^{0.82}$$

wherein Q_m = heat input in kW and A_m = wetted area of vessel in m².   (1) True   (2) False

1-2 : Using the fundamental definitions and units, calculate the conversion factor for 1 std. atm. to bar given the following data: density of Hg = 13.595 g/cm³, g = 9.807 m/s²

1-3: Molar heat capacity of methane is given by $C_P = 3.204 + 18.41 \times 10^{-3}\ T - 4.48 \times 10^{-6}\ T^2$ where C_P is   in cal/(g-mol)(K) and T is in degrees Kelvin. Obtain an expression for the heat capacity when specific
heat is desired in units of kJ/(kg-°C) and temperature in degrees °C.

Solutions to Example Problems:

1-1:

In the following answers, it is not necessary to write the explanations given in braces after the answer if it were an actual exam. They are given here to explain the reasoning behind the answer as a review.)

   a.   False. (British Engineering system uses Time, Length and Force as fundamental units. while
           mass is a derived unit in this system)

   b.   False.(Dimnsional inconsistency of an equation shows that it is wrong but the consistency is
           not enough to prove its accuracy because in the analysis performed, all the variables involved
           may not have been accounted for).

   c.   True (The operation of using dimensionless conversion factors is equivalent to multiplication
               by unity.)

   d. True. (Its symbol is cd.)

   e. True. (The dimensions on the left of the equation are L² .and same is true for right hand side.).

   f.   True. (Using Newtons's second law, F = ma and dividing both sides of the force equation by
           V - Volume of the substance , one obtains F/V = (m./V) a which can be written as $w = \rho g$
           when a = g, where ρ is the density of the substance in kg /m³ and g is the local acceleration of
           gravity.)

g. True. ($KE = mu^2/2 = (10\ kg)(3^2\ m^2/s^2)/2 = 45\ kg\text{-}m^2/s^2 = 0.045\ kJ$)

h. True. (The pressure gauge indicates gage pressure. By adding local barometric pressure $(750.3/760) \times 1.013 = 1.$ bar to it, the pressure 5 bar in absolute units is obtained.)

i. True. (ppm means one part in one million. by weight. 8 ppm means 8 parts per million by weight This is equivalent to a weight fraction of 0.000008. ρ of water 1000 kg/m^3. Hence oxygen in water = 455(1000) x 0.000008 = 3.64 kg/h).

j. True. [1 ft^2 = (1/10.765) m^2] and 1 Btu/h = 0.293x10^{-3} kW. When these conversion factors are substituted in the equation using British units, and numerical values are combined, the constant in SI units results).

1-2 :

$$1\ std\ atm = \rho g \Delta z = 13.595\ \frac{g}{cm^3} \times 9.807 m/s^2 \times 760\ mmHg$$

$$= \frac{13.595}{1000\ \frac{g}{kg}}\ \frac{g}{cm^3 \times \frac{10^{-6}m^3}{cm^3}} \times 9.807 m/s^2 \times 760\ mm \times \frac{10^{-3}}{mm}$$

$$= 101328\ \frac{kg}{m.s^2}$$

$$= \mathbf{101328\ Pa} = \mathbf{101.3\ kPa}$$

$$= \mathbf{101.3/100} = \mathbf{1.013\ bar}$$

1-3:

1 kg of CH$_4$ = 1/16 kg-mol CH$_4$
1 cal = 4.184 J;   1 ^{0}C = 1 K ;
T = t$_c$ + 273.16 ^{0}C

Molar heat capacity $C_P = M\ c_p \left(\frac{cal}{g\text{-}K}\right) = Mc_p \left(\frac{4.184\ J}{\frac{g}{1000\ g/kg}K}\right) = 16c_p \left(\frac{4.184 \times 1000\ J}{kg.K}\right)$

$$= \frac{66.944\ kJ}{kg\text{-}K}\ c_p$$

$c_P \left(\frac{kJ}{kg\text{-}K}\right) = \frac{C_p}{66.944}$

$$= \frac{1}{66.944}\ [\ 3.204 + 18.41 \times 10^{-3}\ (t_c + 273.16) - 4.48 \times 10^{-6}\ (t_c + 273.16)^2\ ]$$

$$= \frac{1}{66.944}\ [\ 7.9 + 1.6 \times 10^{-2}\ t_c - 4.48 \times 10^{-6}\ t_c^2\ ]$$

$$= 0.118 + 2.39 \times 10^{-4}\ t_c - 6.7 \times 10^{-8} t_c^2$$

Where t$_c$ is in ^{0}C.

References:
1. Das D. K., and R.K. Prabhudesai; **Chemical Engineering License Review**, Engineering Press, 2nd ed. (1996)
2. **Fundamentals of Engineering Reference Handbook**, NCEES (1994)
3. Noble R.D, Material and Energy Balances, **AIChE Modular Instruction** , vol 1, Series F (1981)
4. Perry R.H.,.**Chem. Engineers' Handbook**, McGraw-Hill Book Co.; 6th ed (1984), pp 1-1 to 1-18

CHAPTER 2
MATERIAL AND ENERGY BALANCES

ACCOUNTING PRINCIPLE : This can be expressed by the equation:

$$Q_E - Q_B = \Sigma\, Q_I - \Sigma\, Q_O + \Sigma Q_P$$

where   Q = physical quantity chosen for accounting
Q_E = the quantity at end of accounting period
Q_B = the quantity at the beginning of the accounting period
Q_I = the quantity which enters the system during accounting period
Q_O = the quantity which leaves the system during accounting period
Q_P = Quantity created or destroyed during accounting period

In ordinary processes involving chemical (not nuclear) reactions, $\Sigma Q_P = 0$ If the quantity in the system does not change with time , $Q_E - Q_B = 0$ and the accounting equation reduces to $\Sigma Q_I - \Sigma Q_O = 0$. In this case, input equals output. The system is said to be under <u>steady state condition</u>. The accounting principle can also be used to make an energy balance for a system. In carrying out mass and energy balances, several definitions and relations about different properties of substances are useful. These are:

Density of a substance, ρ = mass/volume   and   Specific gravity of a substance,
$s = \rho/\rho_r$
where ρ_r is the density of the reference substance. For liquids and solids, water is usually the reference substance at 4 °C(ρ = 1000 kg/m³)
Other specific gravity conventions are in use. For example, in petroleum industry, API gravity scale is used and is defined as $^0API = 141.5/s - 131.5$
Other sp. gravity scales include Baumè, Brix, and Twadell.   Specific volume = $\underline{V} = 1/\rho$

Mass fraction = (mass of component A)/(Total mass of all components present)
Mass % = mass fraction x 100

Mole fraction x_A = (Moles of component A)/(Total moles of all components present)
where x_A = mole fraction of component
Mole fraction in vapor phase is denoted by y, and that in liquid is usually denoted by x.
Mole % = mole fraction x 100.

Concentration of a species in a mixture is defined as its quantity per unit volume of the mixture.

Volume concentration of A = V_a / V_M
Where V_a = Volume of component A, and V_M = Volume of mixture
Both V_A and V_M are at the temperature and pressure of the mixture.

Mass concentration of a species in a mixture = (Mass of species I)/Volume of mixture
Average molecular weight of a gas mixture $M_{av} = \Sigma x_i M_i$
where x_i and M_i are the mol fraction and molecular weight of component i.

Molar concentration of a species i in a mixture = (Moles of species I)/Volume of mixture
Molarity = M = (g-moles of species i in a mixture)/1 liter of solution
Normality = N = (gram-equivalent weights of species i)/1 liter of solution
Molality = m = (g-moles of species /1 kg solvent

Flow rate is the amount of material that passes a given reference point in a system per unit of time.
Mass flow rate and volumetric flow rate are the most important methods of expressing flow rates.
VAPOR PRESSURE : The pressure exerted by a vapor in equilibrium with its liquid at a given temperature is termed its vapor pressure. If the gas phase consists of more than one component, the total pressure exerted on the liquid surface is the sum of the pressures exerted by the various components. The pressure contribution by a component to the total pressure is termed its partial pressure.

At saturation, component partial pressure = component vapor pressure
The temperature at which a vapor is saturated is called the **dew point or saturation temperature.**
The **boiling point** of a liquid at a given pressure is defined as the temperature at which its equilibrium vapor pressure equals the total pressure on its surface. The temperature at which a liquid boils under a total pressure of 1.0 atm is called its **normal boiling point**.

The vapor pressure of a pure substance is a unique property of that substance and increases with temperature. Two frequently used relations for the vapor pressure are as follows

Two constant equation: $\ln p = A + B/T$
Where   p = vapor pressure and T = absolute temperature, K

Antoine equation: $\ln p = A - [B/(T + C)]$
Where A, B, and C are constants and T is in degrees K. An important equation relating vapor pressure, heat of vaporization , and temperature is Clausius-Clapeyron equation :

$$\frac{dp}{p} = \frac{\Delta H_v dT}{RT^2}$$

where R = Gas-law constant and ΔH_v = heat of vaporization..
Watson's empirical equation correlating heat of vaporization and temperature is

$$\frac{\lambda_2}{\lambda_1} = \left(\frac{1 - T_{r2}}{1 - T_{r1}} \right)^{0.38}$$

where λ_2 = Heat of vaporization cal/gmol at temperature T_2
         λ_1 = Heat of vaporization cal/gmol at temperature T_1
         T_{r2} = Reduced temperature at T_2 and T_{r1} = Reduced temperature at T_1

IDEAL GASES:
    Boyle's Law- PV = constant at constant temperature
    Charle's law- P/T = constant   at constant volume
    Equation of state $PV = nRT$ where R is gas law constant and n = number of moles

IDEAL GAS MIXTURES

Dalton's Law: $\qquad P = p_A + p_B + p_C + \cdots$

Amagat's Law: $\qquad V = V_A + V_B + V_C + \cdots$

$$p_A = y_A P$$
$$V_A = y_A V$$

Ideal gas law is applicable at low pressures and high temperatures.

Normal molar volume = 22.4 m³/kmol at 0° C and 1.0 atm (760 mmHG).

Normal molar volume = 359 ft³ /lb-mol of an ideal gas at 32° F and 1.0 atm

1 lb-mol of an ideal gas occupies a volume of 379 ft³ at 60 ° F and 1 atm. pressure.

CRITICAL PROPERTIES

Critical temperature, T_c - Temperature at which the molecular kinetic energy of translation equals the maximum potential energy of attraction.

Critical pressure, P_c - Pressure required to liquefy a gas at its critical temperature.

Critical volume, V_c - Volume of gas at critical state (at P_c and T_c)

Reduced Conditions - Reduced temperature, $T_r = T/T_c$

$\qquad\qquad\qquad\qquad$ Reduced pressure , $P_r = P/P_c$

$\qquad\qquad\qquad\qquad$ Reduced volume , $V_r = V/V_c$

Corresponding state

The state of equal reduced temperature and pressure. Many properties of gases and liquids are nearly the same at corresponding state.

NONIDEAL GASES:

$\qquad\qquad$ Compressibility of an actual gas , $z = PV/RT$

$\qquad\qquad$ Critical compressibility, $\qquad z_c = P_c V_C/RT_C$

EQUATIONS OF STATE FOR NONIDEAL I.E. REAL GASES

Van der Waals' Equation : $\quad (P + a/v^2)(v - b) = RT$,

$\qquad$ where $\quad n = 1, \; v_c = 3b$, $T_c = 8a/27Rb$, $P_c = a/27b^2$

Redlich-Kwong Equation : $\quad P = RT/(v - b) - a/\left[T^{0.5}v(v + b)\right]$

$\qquad$ where $a = 0.42748R^2 \, T_c^{2.5}/P_c$ $\quad$ and $\qquad b = 0.08664 \, R \, T_c /P_c$

Virial Equation :

$$\frac{Pv}{RT} = 1 + \frac{B(T)}{v} + \frac{C(T)}{v^2} + \frac{D(T)}{v^3} + \cdots$$

SOLUTIONS

For substances in dilute solution, Raoult's law applies. It states that the equilibrium vapor pressure exerted by a component in a solution is proportional to its mol fraction in the solution. Thus

$$p_A = x_A P_A$$

where p_A = vapor pressure of component A in solution with components B, C, . . .

x_A = mole fraction of A in solution

P_A = vapor pressure of pure component A at the temperature of the solution

Relative vapor pressure

A modified form of Raoult's law is many times used to obtain vapor pressure data for a component from a single experimental point. This is given by

$$p = k\, p_s$$

where   p = vapor pressure of solution., p_s = vapor pressure of pure solvent

k = a proportionality factor called the relative vapor pressure of the solution

Boiling point elevation - Increase in boiling point of a solution compared to pure solvent due the presence of a non - volatile solute in a solvent.

$$\text{B.P. elevation} = \frac{RT^2}{\Delta H_v} \frac{M_s}{1000} \times \frac{w}{M_a}$$

where T - boiling point of pure solvent, K

M_s = molecular wt. of solvent, M_a = molecular wt. of solute

ΔH_v = Molar heat of vaporization of the solvent, cal/g-mol

and   w = grams of solute in 1000 grams of solvent

Freezing point depression - lowering of the freezing point of a solution compared to pure solvent due the presence of a nonvolatile solute.

$$\text{Freezing point depression} = \frac{RT^2}{\Delta H_f} \frac{M_s}{1000} \times \frac{w}{M_a}$$

where ΔH_f = latent heat of fusion, cal/g-mol.

HUMIDITY AND SATURATION

Pure component volume in a saturated gas obeying ideal gas law is given by

$$V_v = V\,(p_v\,/p)$$

where V_v = pure component volume of vapor,

p_v = partial pressure of vapor and p = total pressure

For unsaturated gas mixtures, partial pressure of vapor is less than the vapor pressure.

Relative saturation in % = (p_v/p_s) x 100 = y_r

Percent saturation = (n_v/n_s) x 100 = y_p

where n_v = moles of vapor per mole of vapor-free gas actually present

n_s = moles of vapor per mole of vapor-free gas when the mixture is saturated.

Applying Dalton's Law, $y_p = y_r \left(\dfrac{p - p_s}{p - p_v} \right)$

WATER VAPOR- GAS MIXTURES

Humidity, H = weight of water vapor per unit weight of moisture-free gas.

Molal humidity, H_m = number of moles of water vapor per mole of moisture-free gas

% Humidity = % saturation when applied to water.

Wet-bulb temperature
This is the equilibrium temperature attained by a liquid vaporizing into a gas under adiabatic conditions.

Dry-bulb temperature
This is the temperature of the gas into which evaporation takes place.

Psychrometry is the application of wet and dry-bulb thermometry to air-water system. Humidity chart for water is constructed usually for a total pressure of 1 atm.

SOLUBILITY AND CRYSTALLIZATION
Solubility is the concentration of the solute in a saturated solution at the temperature of the solution. Solutes are of two types. Solutes which form compounds with the solvent and those which do not. Solvates are compounds of definite proportions between solute and solvent. If water is the solvent, the compound is termed hydrate. Solubility diagrams are available for a number of solute-solvent pairs in literature. Material balance problems involving dissolution of solute in a solvent or its crystallization from the solvent can be solved using solubility diagrams. In solvent extraction, the initial solvent from which a soluble component is to be removed is called raffinate solvent and the immiscible solvent used for extracting the solute is termed extract solvent.
If there exists an equilibrium between the raffinate and the extract phases, the distribution coefficient for the solute is given by

$$K = \frac{C_E}{C_R}$$

where   K = distribution coefficient, C_E = solute concentration in extract phase
    and C_R = solute concentration in raffinate phase.

ENERGY BALANCES
Potential energy, PE- external energy possessed by a system due to its position in a gravitational field.

Kinetic energy, KE - All energy of a system associated with the macroscopic motion of its mass.
$= M(u^2/2g_c)$

Internal energy, U - total energy possessed by a system due to its component molecules, their relative positions and their movement. For a closed system (constant mass), it is a state property.

Heat energy, Q - energy transferred across the boundaries of a system because of temperature difference.

Work, W- work is defined as W= force x distance It is a form of energy transferred across the boundaries of a system and is not system state property.

Enthalpy, H -   This is defined as H = U + PV

Both U and H are determined relative to a selected reference state. The reference state of zero enthalpy is chosen as 1 atm and 0 °C. For water, it is taken as 0 °C and its own vapor pressure at 0 °C.

The accounting principle can be applied to energy balance. which can be written as follows

$$(U + PE + KE)_E + (U + PE + KE)_B = \Sigma(H + PE + KE)_I - \Sigma(H + PE + KE)_o + Q - W$$

where
$(U + PE + KE)_E$ = sum of internal, potential and kinetic energies at the end of accounting period.
$(U + PE + KE)_B$ = sum of internal, potential and kinetic energies at the beginning of accounting period.
$(H + PE + KE)_I$ = Energy transfeered into the system, that is associated with mass transfer
$(H + PE + KE)_o$ = Energy transfeered out of the system , that is associated with mass transfer
Q = heat transferred across system boundaries during accounting period
W = all work that crosses system boundaries during the accounting period.

For nonflow reaction taking place at constant pressure,
$$Q = \Delta H$$
For a flow reaction with negligible change in potential and kinetic energies and with no work, the energy balance also reduces to $Q = \Delta H$
For a nonflow reaction taking place at constant volume, the energy balance reduces to $Q = \Delta U$

Heat of chemical reaction = Change in enthalpy of the reaction system at constant pressure. This is a function of the nature of reactants and products involved and also their physical states.

Standard heat of reaction = change of enthalpy when the reaction takes place at 1 atm pressure in such manner that it starts and ends with all involved materials at a constant temperature of 25 °C. In exothermic reactions, heat is evolved. In endothermic reactions, heat is absorbed.

Heat of formation of a chemical compound is the change in enthalpy when elements combine to form the compound which is the only reaction product. Standard heat of formation of a compound , ΔH_f is the heat of reaction at 25 °C.

LAWS OF THERMOCHEMISTRY :
(1) At a given temperature and pressure, heat of formation of a compound from its elements is equal to the heat required to decompose the compound into its elements. (2) The total change in enthalpy of a system depends only on the temperature, pressure, state of aggregation and state of combination at the beginning and at the end of the reaction and is independent of the number of intermediate chemical reactions involved. This principle is also known as the Law of Hess. This principle is used to calculate the heats of formation of compounds knowing the heats of intermediate reactions For example, the following three reactions allow to obtain the heat of formation of ethane.

$C_2H_6 (g) + 3\frac{1}{2} O_2(g) = 2CO_2(g) + 3H_2O(l)$ $\Delta H_1 = - 372.82$ kcal/g-mole (a)

$2C (\beta) + 2 O_2(g) = 2 CO_2(g)$ $\Delta H_2 = - 188.1$ kcal/g-mole (b)

$3H_2 (g) + 1\frac{1}{2} O_2 (g) = 3H_2O(l)$ $\Delta H_3 = - 204.95$ kcal/g-mole (c)

Equation b + Equation c - Equation a gives

$2C (\beta) + 3H_2 (g) = C_2H_6 (g)$

and ΔH_f of ethane = - 188.1 - 204.95 - (- 372.82) = - 20.23 kcal/g-mole

Standard heat of combustion, ΔH_c = the enthalpy change resulting from the combustion of a substance in its normal state at 25 °C and atmospheric pressure, with the combustion beginning and ending at 25 °C. Generally gaseous CO_2 and liquid water are the combustion products. It is possible to calculate the heat of formation of a compound from its heat of combustion provided the heats of formation of other substances which participate in the reaction are known.

Standard heat of reaction, ΔH_R can be calculated from the heats of formation of reactants and the products using the following relation

$$\left[\Delta H_{reaction} = \Sigma \Delta H_{f(products)} - \Sigma \Delta H_{f(reactants)} \right] \text{ at } 25\,^0C$$

It can also be calculated from the heats of combustion of the reactants and products by the relation

$$\left[\Delta H_R = \Sigma \Delta H_{c(reac\tan ts)} - \Sigma \Delta H_{c(products)} \right] \text{ at } 25\,^0C$$

(For more review of energy balances, see review of thermodynamics in a later review devoted to it.)

FUELS AND COMBUSTION
Heating value of fuel is its standard heat of combustion numerically but of opposite sign. Two types of heating values are defined . Total heating value is the heat evolved in the complete combustion of a fuel under constant pressure at temperature of 25 °C with all the water initially present and that produced in the combustion are condensed to the liquid state at 25 °C. This is also termed as higher or gross heating value and is denoted as HHV. Net heating value also termed LHV or lower heating value is defined in a similar manner but the final state of the water is taken as vapor at 25 °C.

COAL ANALYSIS
<u>Ultimate analysis</u> involves the determination of all elements in the sample. e.g. carbon, hydrogen, sulfur etc.

<u>Proximate analysis</u> involves determination of four groups namely fixed carbon, volatile matter, ash, and moisture.

Total heating value of coal , H.V. = 14490C + 61000 H_a + 5550S (This is Dulong's formula) where C, H_a, S = wt. fractions of carbon, available hydrogen, and sulfur respectively and H.V. = heating value of coal, Btu/lb

Net H.V. of coal = Total H.V. - 8.94 x H x1050
Net H.V. = heating value, Btu/lb and H= wt fraction of total hydrogen including available hydrogen, the hydrogen in moisture, and hydrogen in combined water.

Petroleum fuels: U.O.P characterization factor is given by $K = \dfrac{\sqrt[3]{T_B}}{s}$

where K = U.O.P characterization factor,

     T_B = average boiling point, $^{\circ}R$ at 1 atm.

     s = specific gravity at $60^{\circ}F$

EXAMPLE PROBLEMS :

2-1: A fuel gas has the following volumetric composition. Assume ideal gas behavior.

        Methane   85% ,   Ethane 10.5 % , Nitrogen 4.5%

Calculate   a. composition of the gas in mole %.

             b. composition in wt%.

             c. Average molecular wt of the gas.

             d. Density of gas at normal conditions in (1) kg/m^3 (2) lb/ft^3

             e. What is the partial pressure of nitrogen in kPa if the total pressure is 1.013 barG, and barometric pressure is 1.013 barA.

2-2: 2000 kg/h of a mixture consisting of 60 wt.% benzene and 40 wt% tolune is to be separated in a distillation column. The distillate is to contain 98 wt% benzene and 95 % of feed benzene is to be recovered as distillate. what are the flow rates of the distillate and bottoms product and the composition of the bottoms product.

2-3: 100 kg of a mixture of sodium sulfate crystals ($Na_2 SO_4 \ 10H_2 O$) and sodium chloride (NaCl) is heated to drive away all the water. Final wt of the dry mixture is 58.075 kg. Calculate
(a) mass of soduim sulfate crystal and sodium chloride in the original mixture.
(b) molar ratio of dry $Na_2 SO_4$ and NaCl in the original mixture.

2-4: A solution of $NH_4 Cl$ is saturated at 70 $^{\circ}C$. Calculate the temperature to which this solution must be cooled in order to crystallize out 45% of the $NH_4 Cl$. The solubilities of $NH_4 Cl$ in water are :

Temperature °C	70	10	0
Solubility g/100 g of water	60.2	33.3	29.4

2-5: Calculate the amount of $H_2 S$ in cubic meters measured at 49 °C and at a pressure of 0.2 barG, which may be produced from 10 kg of FeS.

2-6: Determine the change in enthalpy in kJ when 64.40 kg of $Na_2 SO_4.10H_2O$ is dissolved in 108 kg of water at 25 $^{\circ}C$

Species	Heat of formation at 25°C, kJ/kg.mol
Na_2SO_4 (infinite dilution)	-1.385×10^6
$Na_2SO_4.10H_2O$ (infinite deilution)	-4.325×10^6
Water	-0.286×10^6

Molar entrhapy of Na_2SO_4 (40 moles of water = -0.0109×10^6 kJ/kg.mol at 25°C

2-7: A mixture of nitrogen and water vapor contains 25% by volume water vapor. at a temperature of 80°C and 101.3 kPa. pressure. The vapor pressure of water at 80 $^{\circ}C$ = 47.3 kPa.
Calculate (a) relative saturation and (b) % saturation of the mixture.

2-8:A solution of $NaNO_3$ at a temperature of 60 °C contains 45 % $NaNO_3$ by weight.
   (a) What is the percentage saturation of this solution? At 60°C, Solubility of $NaNO_3$ = 124 g/100g of water.
   (b) Calculate the weight of $NaNO_3$ that will be crystallized if 100 g of this solution is cooled to 10°C.
   (c) Calculate the percentage yield of this crystallization process.
   (d) What is the molarity of the original solution? sp. gr. of solution = 1.3371 at 60°C.
   (e) What is the molality of the original solution?
   (f) What is the concentration of $NaNO_3$ in original solution in grams per 100 g of water.

2-9 : Air at 100 kPa pressure and a dry-bulb temperature of 93 °C and wet bulb temperature of 35 °C is fed to a dryer. In the dryer 0.014 kg-mol of water evaporates per kg-mol of air fed to the dryer. If the vaporization of water in the dryer is adiabatic, calculate (a) The dry-bulb temperature (b) the wet-bulb temperature and (c) the percentage saturation of the air leaving the dryer.

2-10: 1000 kg of a solution of 10 % H_2SO_4 at 27 °C is to be fortified to 50 % strength by the addition of 98% H_2SO_4 which is at 21 °C. How much heat is removed by the cooling system if the final temperature is to be 38 °C? The enthalpies are as follows:

10 % H_2SO_4 $\underline{H}$ = 12 Btu/lb   at 27°C = 12x2.326 = 27.912 kJ/kg
98 % H_2SO_4 $\underline{H}$ = -3 Btu/lb   at 21 °C = -3x2.326 = 6.978 kJ/kg
50 % H_2SO_4 $\underline{H}$ = - 84 Btu/lb at 38 °C = -84x2.326 = -195.384 kJ/kg

SOLUTIONS TO EXAMPLE PROBLEMS

2-1:   Basis 100 kmol of fuel gas

   (a) For an ideal gas, vol. % and mol % are the same. Therefore , the molar composition of the gas is: Methane 85 mol%, Ethane 10.5%, and Nitrogen 4.5 mol%

   (b) Composition in wt %

component	mol%	kmols	mol. wt.	wt in kg	wt%
methane	85	85.00	16	1360	75.51
ethane	10.5	10.50	30	315	17.49
Nitrogen	4.5	4.5	28	126	7.00
				1801	100.00

   (c) Mol fractions $y_{methane}$ = 0.85, y_{ethane} = 0.105, and $y_{nitrogen}$ = 0.045
   Average molecular wt of gas = 0.85x16 + 0.105x30 + 0.045x 28 = 18.01

   (d) Density at standard conditions

   (i) 1 kmol occupies 22.4 m^3 at std. conditions of 0 °C and 1 atm = 1.013 bara..
   Therefore, density ρ = mol. wt/vol = (18.01/22.4) = 0.804 kg/m^3 at std. conditions
   (ii) 1 kg = 2.2046 lb mass   1 m^3 = 35.3 ft^3
   density in lb/ft^3 , ρ = (0.804x2.2046)/(1x35.3) = 0.0502 lb_m /ft^3
   (e) Total pressure = 1.013bar G = 1.013 + 1.013 = 2.026 barA = 2.026x100 = 202.6 kPa
   Therefore, partial pressure of nitrogen = 0.045x202.6 = 9.117 kPa

2-2 :

Material Balance

Let F = feed,   B = bottoms product,   D = distillate   all   in   kg/h

Also y = mass fraction benzene in distillate = 0.98

x_B = mass fraction benzene in bottoms product

Basis: 2000 kg/h

Overall balance   F = B + D

Benzene balance $Fx_f = yD + Bx_B$

Also from the given composition of distillate and the recovery of benzene in distillate,

$$0.95\ Fx_f = 0.98\ D$$

By substitution of given values, the equations become

$$B + D = 2000 \dots\dots\dots\dots\dots\dots (1)$$
$$(0.6\ F) = 0.98\ D + Bx_B \dots\dots\dots(2)$$
$$0.95(2000)(0.6.) = 0.98\ D \dots\dots\dots (3)$$

From third equation $D = \dfrac{0.95 \times 2000 \times 0.6}{0.98} = 1163.27\ \text{g/h}$

Substituting value of D in equation 1, B = 836.73 kg/h

Using values of B and D obtained above and substituting in equation 2,

$$0.95(0.6\ F) = 0.98\ (1163.73) + 836.73x_B$$

From which $x_B = \dfrac{0.6(2000) - 0.98(1163.27)}{836.73} = 0.0717$ wt fraction

Therefore , bottoms composition is : benzene = 7.17 wt% ,   toluene = 92.83 wt%

Distillate = 1163.27 kg/h   and bottoms product = 836.73 kg/h

2-3 :

Molecular wts Na_2SO_4 = 142, $Na_2SO_4.10H_2O$ = 322, H_2O = 18,   NaCl = 58.5

$$Na_2SO_4.10H_2O \rightarrow Na_2SO_4 + 10H_2O$$
$$322 \qquad\qquad 142 \qquad\qquad 180$$

let x = kg of $Na_2SO_4.10H_2O$ in the original mixture

y = kg of NaCl in the original mixture

Material balance equations: (Basis of calculation : 100 kg of undried mixture)

1. Original mixture   x + y = 100

2. Dried mixture $x - x\left(\frac{180}{322}\right) + y = 58.075$

Subtracting eq. 2 from eq. 1 gives

$$\left(\frac{180}{322}\right)x = 41.925 \text{ or } x = 41.925 \times (322/180) = 75 \text{ kg}$$
$$y = 100 - 75 = 25 \text{ kg}$$

Dry Na_2SO_4 in mixture = (142/322) x 75 = 33.075 kg = 33.075/142 = 0.2329 kg-mol

NaCl in the mixture = 25/58.5 = 0.42735 kg-mol

Molar ratio of dry Na_2SO_4 to NaCl = 0.2329/0.42735 = 0.545

2-4 :

At 70 $^{\circ}$C, solubility of NH_4Cl = 60.2 g/100 g of water

45 % of NH_4Cl to be crystallized.

NH_4Cl in solution = (1 - 0.45) x 60.2 = 33.1 g

Solubilities of NH_4Cl at 0 and 10 $^{\circ}$C are 29.4 and 33.3 g/100 g of water respectively.

by interpolation, the temperature at which the solubility is 33.1 g/100 g of water is 9.5 $^{\circ}$C.

2-5 :

$$FeS \rightarrow H_2S$$
$$88 \rightarrow 34$$

Amount of H_2S produced = (34/88)x10 = 3.864 kg per 10 kg of FeS

Since no other data are given, assume ideal gas behavior.

Pressure = 0.2 + 1.013 = 1.213 barA. ,   T = 50 + 273 = 323 K

kg-mol of H_2S = 3.864/34 = 0.113647 kg-mol

Volume at std. conditions = 0.113647(22.4) = 2.546 m^3

Volume at 50 $^{\circ}$C and 1.213 bar Abs , V= (2.546) $\left(\frac{323}{273}\right)\left(\frac{1.013}{1.213}\right)$ = 2.516 m^3.

2-6 :

$142+180$

64.4 kg of $Na_2SO_4\ 10H_2O$ = 64.4/322 = 0.2 kg-mol

108 kg of water = 108/18 = 6 kg-mol

Basis of calculation = 1 kg-mol of $Na_2SO_4\ 10H_2O$

Corresponding amount of water is (1/0.2) x 6 = 30 kg-mol of water

Heat of reaction for the reaction Na_2SO_4 + $10H_2O$ $\rightarrow$ $Na_2SO_4 10H_2O$

 - 1.385x10^6 10(- 0.286x10^6)   - 4.325x10^6

ΔH_R = -4.325x10^6 - (-1.38x10^6) - (10x0.286x10^6) = -0.085x10^6 kJ/kgmol

This is molal enthalpy of $Na_2SO_4\ 10H_2O$ relative to Na_2SO_4 and H_2O at 25 $^{\circ}$C.

Molal enthalpy of Na_2SO_4 (n_1 = 40) relative to Na_2SO_4 and H_2O is = -0.0109x10^6

kJ/kgmol at 25 $^{\circ}$C.   Then for the reaction,

$$Na_2SO_4\ 10H_2O + 30\ H_2O \rightarrow Na_2SO_4(\ n_1 = 40)$$

ΔH = - 0.0109x10^6 - (-0.085x10^6) = + 0.0741x10^6 kJ/kgmol

Therefore, change in enthalpy when 0.2 kg-mol of $Na_2SO_4\cdot10H_2O$ is dissolved in 6 lkg-mol

of water ΔH = (0.2) x 0.0741x10^6 = 14820 kJ

2-7:

(a) partial pressure of water vapor = 0.25 (101.3) = 25.33 kPa

vapor pressure of water = 47.3 kPa

Relative saturation = (p_w/p)   100 = (25.33/47.3) x100 = 53.55 %

(b) Basis 1 kg-mol of mixture

Mols of nitrogen in mixture = 1.0 - 0.25 = 0.75 kg-mol

Therefore, % saturation = (0.25/ 0.75) x 100 = 33.33 %

2-8:

(a) At 60 $^{\circ}$ C, solubility of $NaNO_3$ = 124 g / 100 g of water.

Initial solution is 45 % $NaNO_3$ by weight, or (45/55) = 81.82 g/100g of water.

Therefore,   % saturation = 81.82 / 124 = 66 %

(b) Solubility of $NaNO_3$ at 0 $^{\circ}$C = 73 g per 100 g of water or $NaNO_3$ is 42.2 % by

weight. Original solution has 50 g $NaNO_3$.

Let x be the amount of $NaNO_3$ precipitated on cooling

By material balance on $NaNO_3$, $\quad 100 \, (0.45) = x + 0.422(100 - x)$
Which on solving gives x = 4.84 g
(c) % yield = (4.84/45) x 100 = 10.76 %
(d) MW of $NaNO_3$ = 85.
$NaNO_3$ in solution = 50/85 = 0.53 g-mole in 100 g of solution..
specific gravity of 45 % $NaNO_3$ solution = 1.3371 at 60 °C
volume of solution = 100/1.3371 = 74.79 cc
Therefore, molarity = (0.53/74.79) x 1000 = 7.09 g-moles/liter of solution
(e) Molality = (0.53/50) x 1000 = 10.6 gmoles/1000 g of water.
(f) Original solution contains 45 g of $NaNO_3$ in 55 g of water.
Concentration of $NaNO_3$ = (45 /55)x 100 = 81.82 g/100 g of water.

2-9: At 93 °C dry-bulb and 35 °C wet-bulb temperature, molal humidity is
0.02 kg-mol/kg-mol dry air This is read from psycrometric chart given in
Chemical Engineering Licese Review[1].
Water evaporated in dryer = 0.03 kg-mol/kg-mol dry air
Therefore, humidity of air leaving the dryer = 0.02 + 0.03 = 0.05 kg-mol/kg-mol dry air.
Since vaporization is adiabatic, wet-bulb temperature remains the same and equals 35 °C.
Dry-bulb temperature at molal humidity of 0.05 kg-mol/kg-mol dry air and wet-bulb
temperature of 35 °C is read from the same chart and is equal to 47.5 °C .
Saturation humidity at 47.5 °C dry-bulb = 0.12 kg-mol/kg-mol dry air.
Therefore, % saturation = (0.05/0.12)x100 = 41.7 %

2-10: For solution of this problem, use enthalpy composition diagram[2,3] for sulfuric acid and water
system.
Material Balance
Let x be the amount of 98 % acid added. Then balance on H_2SO_4 gives
$\quad$ 0.98 x + 0.1(1000) = 0.5(1000 + x)
Therefore, x = 833.3 kg of 98 % H_2SO_4
Energy balance gives the heat to be removed by the cooling system.
Q = - 195.384 (1000 + 833.3) - 1000(27.912) - 833.3(-6.978) = - 380295 kJ
Heat to be removed by the cooling system = 380295 kJ

References
(1) Das D.K. and R.K. Prabhudesai, **Chemical Engineering License Review**, Engineering Press,
2nd Ed., 1996
(2) Hougen O.A, et al; **Chemical Process Principles,** part 1, John Wiley & Sons,
New York, 1954
(3) Henley E.J., et.al., **Material and Energy Balance Computations**, John Wiley & Sons,
New York, 1969
(4) **Material & Energy Balances**, AIChE Modular Instruction Series, V1-4
(5) Perry R.H. and D.W. Green, **Chemical Engineers' Handbook**, McGraw-Hill,
New York, 6th Ed., (1984)

CHAPTER 3
THERMODYNAMICS

System: This is a portion of the universe (e.g. a substance or a group of substances) set apart for study.

Closed system - A system with constant mass

Open system - A system with variable mass (mass is transferred across the boundaries of the system).

Thermodynamic properties: Any measurable characteristic of a system is called its property. eg temperature, pressure, mass, area volume surface tension etc. **Intensive** property is independent of mass. **Extensive** property depends on the mass of the system. **State** properties depend upon the thermodynamic state of the system and are independent of the path the system takes to reach that state. Some state properties are: pressure P, temperature T, specific volume v, internal energy U, enthalpy H , entropy S , Gibbs free energy G, Helmholz free energy A, heat capacities at constant pressure C_P and at constant volume C_V.

MASS BALANCE:

Application of the accounting principle to mass of a system over an accounting period yields the mass balance equation:

$$M_E - M_B = \Sigma M_I - \Sigma M_O + \Sigma M_P - \Sigma M_C$$

ENERGY BALANCE **First Law of Thermodynamics or principle of conservation of energy:**

Application of accounting principle to energy of a system over a chosen accounting period yields :

$$(U + PE + KE)_E - (U + PE + KE)_B = \Sigma(H + PE + KE)_I - \Sigma(H + PE + KE)_I + Q - W - \Sigma E_C$$

Where U - Internal energy, PE-potential energy, KE-kinetic energy, H-enthalpy, Q-heat transferred across the boundary of the system during accounting period, W- all work that crosses the system boundaries during the accounting period and ΣE_C - Energy conversions due to atomic transmutations. In ordinary non-nuclear reactions or for ordinary physical and chemical processes $\Sigma E_C = 0$. Terms in the energy balance equation are defined in the review of material balances. When Q enters the system, it is taken as positive. when it leaves the system it is taken as negative. Work done by the system is positive while work done on the system is negative.

Work = force x distance. When fluid pressure is the only force acting on the system , $W = \int_{V_1}^{V_2} PdV$

Second Law of Thermodynamics : Heat cannot be converted into work quantitatively and cannot be transferred from a lower temperature to a higher temperature without the aid of an external energy. This principle is covered by second law of thermodynamics which is stated as follows:

No practical engine can convert heat into work quantitatively or it is impossible for a self-acting machine unaided by an external agency to transfer heat from a lower temperature to a higher one.

Carnot principle: Efficiency of a reversible engine depends only on temperature levels at which the heat is absorbed and rejected and is independent of the medium used. Thus

$$\text{Efficiency } \eta_{rev} = \frac{W_{rev}}{Q_1} = \frac{Q_1 - Q_2}{Q_1} = \frac{T_1 - T_2}{T_1}$$

For an engine undergoing reversible Carnot cycle, $\oint \frac{dQ_R}{T} = 0$ This is the change in entropy and it is a state property.

Work done by the engine using a finite temperature difference: $W = Q_1 \left(\frac{T_1 - T_2}{T_1} \right) - T_2 S_p$

Where S_P = entropy of production. $S_p = 0$ for a reversible process. $S_p > 0$ for an irreversible

process.

General statement of second law of thermodynamics :

$$S_E - S_B = \Sigma S_t + \Sigma \frac{Q}{T_B} + S_P - S_C \qquad S_P \geq 0$$

where S_t = Entropy changes due to transfer of mass across the boundary of the system.

S_C = Entropy changes occurring in a system due to atomic transmutations.

For constant mass and composition system undergoing reversible process with pressure as the only external force

$$dU = TdS - PdV$$

Also

$$dH = TdS + VdP$$

Helmholtz function   A = U - TS   and $dA = PdV - SdT$

Gibb's free energy   G = H - TS   and $dG = VdP - SdT$

Calculation of entropy changes:

Energy absorption at constant temperature No chemical reaction); $\Delta S = \frac{Q}{T}$

Isothermal phase change: $\Delta S = \frac{\lambda}{T}$

Heating or cooling without chemical change: $\Delta S = \int_{T_1}^{T_2} \frac{d'q}{T}$

For a constant-pressure condition, when heat capacity is independent of temperature

$$\Delta S = \int_{T_1}^{T_2} \frac{nC_P dT}{T} = nC_P \ln \frac{T_2}{T_1}$$

If the heat capacity varies with the temperature, equation relating heat capacity with temperature is used and the equation is then integrated.

Availability function $B = -[\Delta H - T_0 \Delta S]_{T_1}^{T_0}$ For a flow system, $\Delta B = \left[\Delta H - T_0 \Delta S \right]_{T_1}^{T_2}$

When KE and PE changes occur, for a flow system $\Delta B = \left[\Delta H - T_0 \Delta S + \frac{\Delta u^2}{2g_c} + \frac{g}{g_c} \Delta Z \right]_{T_1}^{T_2}$

PROPERTIES OF IDEAL GASES

For ideal gas $PV = nRT$

P-pressure , T-temperature, V-volume, n-number of moles, and R-gas law constant

Boyle's law $P_1 V_1 = P_2 V$ at constant T and n

Charles' law $\frac{V_1}{T_1} = \frac{V_2}{T_2}$ at constant P and N.

Dalton's law of partial pressures and **Amagat's law** of partial volumes also apply to ideal gases.

Thermodynamic relations for an ideal gas

Constant volume (isometric) process: $dU = dQ = C_v dT$ or $\Delta U = Q = \int C_v dT$

Constant pressure (isobaric) process: $dH = dQ = C_P dT$ or $\Delta H = Q = \int C_P d$

Constant temperature (isothermal) process $Q = W = RT \ln \frac{V_2}{V_1} = RT \lg \frac{P_1}{P_2}$

Adiabatic (dQ = 0) process: $\frac{T_2}{T_1} = \left(\frac{V_2}{V_1} \right)^{k-1} = \left(\frac{P_2}{P_1} \right)^{(k-1)/k}$ or $P V^k = $ constant.

where k = ratio of specific heats = $\frac{C_p}{C_v}$

Work for adiabatic process $W = \frac{RT_1}{k-1}\left[1 - \left(\frac{P_2}{P_1}\right)^{(k-1)/k}\right]$

Polytropic compression of an ideal gas: $\frac{n-1}{n} = \frac{k-1}{\eta pk}$ $\quad \eta P$ = polytropic efficiency

Then $\qquad W_P = \frac{RT_1}{n-1}\left[1 - \left(\frac{P_2}{P_1}\right)^{(n-1)/n}\right]$

Also $\qquad \frac{T_2}{T_1} = \left(\frac{P_2}{P_1}\right)^{(n-1)/n} = \left(\frac{V_1}{V_2}\right)^{n-1}$

Entropy changes for an ideal gas: When an ideal gas is compressed from P_1, T_1, V_1 to P_2, T_2, V_2, the entropy change is given by

$$\Delta S_T = \Delta S_1 + \Delta S_2 = \int_{T_1}^{T_2} \frac{C_p}{T}dT - R\ln\frac{P_2}{P_1}$$

(Change temperature at constant pressure and then change the pressure reversibly and isothermally).

By changing temperature at constant volume and then changing volume reversibly and

isothermally,

$$\Delta S_T = \int_{T_1}^{T_2} \frac{C_v}{T}dT + R\lg\frac{V_2}{V_1}$$

PROPERTIES OF REAL FLUIDS

For real fluids, ideal gas law does not hold . Therefore other equations of state are proposed.

These are:

Van der Waal's equation: $\left(P + \frac{a}{V^2}\right)\left(V - B\right) = RT$ a and b are Van der Waal's constants

In terms of critical constants, $a = \frac{27R^2 T_c^2}{64P_c} = (9/8)RT_c V_c$ and $b = \frac{RT_c}{8P_c} = \frac{V_c}{3}$

where b = Excluded volume/mole

Redlich-Kwong equation :

$$P = \frac{RT}{V-b} - \frac{a}{T^{1/2}V(\bar{V}+b)} \qquad \text{Where} \quad a = \frac{0.4278R^2 T_c^2}{P_c} \quad \text{and} \quad b = 0.0867\frac{RT_c}{P_c}$$

Van der Waal's equation in terms of reduced conditions : $\left(P_r + \frac{3}{V_r}\right)\left(V_r - \frac{1}{3}\right) = \frac{8}{3}T_r$

Beattie-Bridgeman Equation of state:

$$Pv^2 = RT\left[v + B_0\left(1 - \frac{b}{v}\right)\right]\left(1 - \frac{c}{vT^3}\right) - A_0\left(1 - \frac{a}{v}\right)$$

The constants are empirically determined.

Reduced conditions : $T_r = T/T_c$, $\quad P_r = P/P_c$, and $V_r = V/V_c$

Compressibility factor:
Equation of state is written as PV = ZnRT where Z is compressibility factor. It is a function of pressure, temperature and the nature of the gas. The **theorem of corresponding states** postulates that all pure gases have the same compressibility factors when measured at the same reduced conditions of temperature and pressure. Generalized Z charts have been developed for the compressibility factors.

Critical compressibility factor $= Z_c = \frac{P_c V_c}{RT_c}$

Virial equation of state: $Z = \frac{PV}{RT} = 1 + \frac{B}{V} + \frac{C}{V^2} + \cdots$

Other form of virial equation is : $Z = \frac{PV}{RT} = 1 + B'P + C'P^2 + d'P^3 +$

For ideal gas, $\qquad dG_T = V\, dP = RT\frac{dP}{P} = RTd\ln P$ at constant temperature.

For non-ideal substance, fugacity is defined by: $dG_{Ti} = RTd\ln f_i$ and $\lim_{p\to 0}\left(f_i/p\right) = 1.0$

For an ideal gas, this becomes $\quad dG_T = V\, dP = RT\frac{dP}{P} = RTd\ln P \qquad$ at constant temperature.

Clapeyron equation:

$$\frac{dp}{dT} = \frac{\Delta H_v}{T\Delta V}$$

where $\Delta H_v=$ heat of vaporization and $\Delta V=$ volume change because of phase change

Third Law of Thermodynamics:
Nerst and Plank postulated that at absolute zero temperature, the absolute entropy of a purely crystalline substance is zero. The absolute entropy of a pure substance at constant pressure can be calculated from:

$$S = \int_0^T \frac{C_{ps}}{T}dT + \frac{\Delta H_f}{T_f} + \int_{T_f}^{T_b}\frac{C_{pl}}{T}dT + \frac{\Delta H_v}{T_b} + \int_{T_b}^{T}\frac{C_{pg}}{T}dT$$

POWER CYCLES : Power cycle involves a series of operations repeated in the same order on the working fluid. Objective is to convert heat into work or work into heat.
The Carnot cycle is depicted on PV and TS diagrams in Figure 3-1.

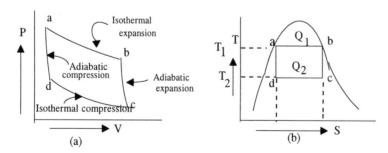

Figure 3-1: Carnot cycle on (a) PV diagram, (b) TS diagram

All steps in a Carnot cycle are reversible. Carnot reversible cycle is a hypothetical heat engine which uses a perfect gas as the working fluid.

Net work is given by $\qquad W_{rev} = \Sigma \oint P dV = Q_1 - Q_2 \quad$ and $\qquad \Delta S_T = \Delta S_1 + \Delta S_2 = 0$

Therefore, thermodynamic ideal efficiency of Carnot cycle $= \dfrac{Q_1 - Q_2}{Q_1} = \dfrac{T_1 - T_2}{T_1} = \dfrac{W_{net}}{Q_1}$

Refrigeration:

This is a process in which a working fluid is used to abstract heat from a low temperature source and to reject it at a higher temperature. A ton of refrigeration is the heat removal rate of 200 Btu/min, 2000 Btu/h, or 288000 Btu/d. One ton of refrigeration is also equal to the latent heat of melting of one ton of ice in 24 hours.

Coefficient of performance $\beta = \dfrac{heat\ removed\ from\ low\ temperature\ region}{work\ of\ compression}$

Reversed Carnot Cycle: The ideal refrigeration cycle is a reversed Carnot cycle It is shown in Figure 3-2 .

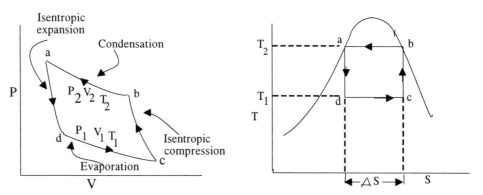

Figure 3-2: Carnot reversed (refrigeration) cycle on (a) PV diagram (b) TS diagram

Coefficient of performance $\beta = \dfrac{heat\ abstracted\ in\ evaporator}{work\ done\ on\ the\ fluid} = \dfrac{Q_1}{-W} = \dfrac{Q_1}{Q_2 - Q_1} \dfrac{T_1 \Delta S}{(T_2 - T_1)\Delta S} = \dfrac{T_1}{T_2 - T}$

Vapor compression with turbine expansion cycle :This is shown in Figure 3-3 .

Coefficient of performance, $\quad \beta = \dfrac{H_g - H_{gl}}{\left(H_d - H_l\right) - \left(H_g - H_{gl}\right)}$

Vapor Compression with Free Expansion: This refrigeration cycle is shown in Figure 3-4

Net refrigeration effect $= \underline{H}_g - \underline{H}_{gl} = \underline{H}_g - \underline{H}_l$

Mass flow rate of refrigerant $= W = \dfrac{12000 \; Btu/h.ton}{Q_1 \; Btu/\,lb}$

Heat of compression $= \underline{H}_d - \underline{H}_g$

Work of compression $= - W_c = (\underline{H}_d - \underline{H}_g)\,W$

Condenser heat load $= Q_2 = \underline{H}_d - \underline{H}_l$

Coefficient of performance $\beta = \dfrac{H_g - H_l}{H_d - H_g}$

hp $= 42.4$ Btu/min

hp/ton $= \dfrac{work\ of\ compression\ (Btu/(min \cdot ton)}{42.4\ Btu/\,min} = \dfrac{4.713}{\beta}$

Rankine Cycle: This is shown in figure 3-5. The efficiency of rankine cycle is given by

$$\eta = \frac{\left(H_c - H_d\right) - \left(H_b - H_a\right)}{H_c - H_b}$$

PROPERTY CHARTS: (1) Pressure-volume (PV) chart with T as parameter (2) Temperature-entropy (T-S) diagram (3) Pressure-enthalpy (PH) diagram (4) Enthalpy-entropy(H-S), also called Mollier diagram. Mollier diagram - more useful in solving the turbine and compressor problems. P-H diagram - useful in solving refrigeration problems. T-S diagram - useful in analysis of engines.

Heats of mixing and enthalpy of a solution:
Heat of mixing = Enthalpy change on dissolving a solute in a solvent. Standard integral heat of solution is change in enthalpy of the system when 1 mol of a solute is mixed with n_1 moles of a solvent at constant temperature of 25 C and 1 atmosphere. The enthalpy is given by

$$H_s = n_1 \bar{H}_1 + n_2 \bar{H}_2 + n_2 \Delta \bar{H}_{s2}$$

Where H_s = enthalpy of $n_1 + n_2$ moles of solution of components 1 and 2 at temperature T relative to temperature T_{ref} (reference temperature). $\bar{H}_1$, $\bar{H}_2$ = molal enthalpies of components 1 and 2 at temperature T relative to the reference temperature T_{ref}, and $\Delta \bar{H}_{s2}$ = integral heat of solution of component 2 at temperature T

Gibbs phase rule: The number of degrees of freedom of a system is given by $F = N - r + 2 - P$ where F - degrees of freedom, N - number of components, r - number of reactions, P - number of phases present.

Fugacity and fugacity coefficient : **Fugacity** of a pure substance is defined by the equation
$$(dG)_T = RT \, d \ln f_i \quad \text{at constant temperature}$$
For ideal gas, this equation becomes
$$(dG)_T = V\,dP = RT\frac{dp}{P} = RTd\ln P$$

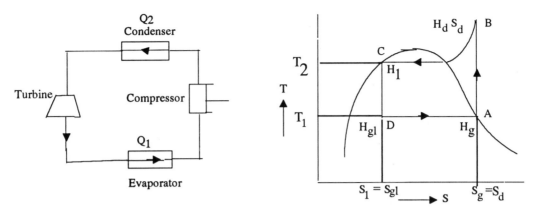

Figure thermo 1-3: Vapor compression refrigeration cycle with turbine expansion
(a) Sketch of system (b) TS diagram

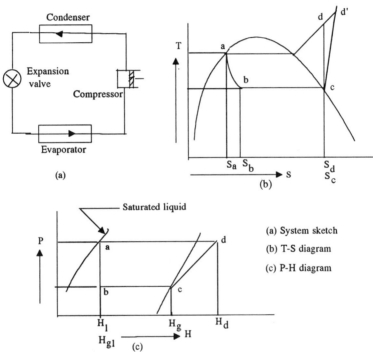

(a) System sketch

(b) T-S diagram

(c) P-H diagram

Figure 3-4: Vapor compression refrigeration cycle with free expansion

A further restriction on the definition of f_i is $\lim\limits_{P \to 0} \dfrac{f}{P} = 0$ The ratio $\dfrac{f}{P}$ is called fugacity coefficient.

Example Problems:

3-1: A piston cylinder initially contains 100 g-mols of an ideal gas at a pressure of 516.3 kN/m²A and 10°C. External pressure is 101.3 kN/m². The piston is weighted with 100 kg of weight and held in place with latches. The temperature is kept constant. When latches are released , the piston moves up and comes to rest when the forces are balanced. Calculate the work done during this expansion. Next the mass is removed and the cylinder allowed to come to equilibrium against the external pressure. What is the work done in this step.? What would be the work done if the gas

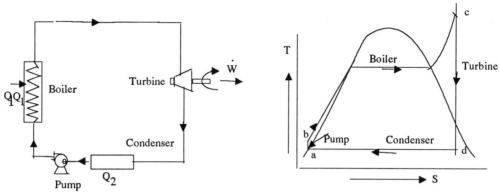

Figure 3-5 : Rankine cycle without superheating   (a) Rankine cycle   (b) T-S diagram

were expanded reversibly against the external pressure. Area of cross section of piston is 0.0029 m². Assume temperature is constant during both expansions. Report calculated work in units of kJ.

3-2: Two Carnot engines are operating in series. The first one absorbs heat at a temperature of 1111 K and rejects heat to the second engine at a temperature T. The second engine receives the heat at the the intermediate temperature T and rejects it to a reservoir at 300 K . Calculate T if (a) the efficiencies of the two engines are equal and (b) the works done by the two engines are equal.

3-3: Express Van der Waal's equation of state as Virial equation.

3-4: Calculate the specific volume of CH_3Cl at 1379 kPa and 205 °C Using Van der Waal's equation . Van der Waal's constants for CH_3Cl are a = 757.4 kPa(m³/kgmol)²   and b = 0.065 m³/kgmol.

3-5: A Carnot engine operates in a closed system by absorbing heat at 927 °C and rejecting heat at 27 °C and produces 5021 kJ of net work. Determine the heat input and output of the engine.

3-6: The constant pressure specific heat of acetonitrile vapor at low pressures is given by the equation
$$C_p = 21.3 + 11.562 \times 10^{-2} T - 3.812 \times 10^{-5} T^2$$
Where T is in K and C_P is in kJ/kgmol-K Estimate specific heat ratio (C_P/C_V) for acetonitrile at 1000 K

3-7: 25 kg of carbon dioxide are to be heated from 300 to 700 K at constant volume. Calculate the number of kJ to be supplied. Specific heat of carbon dioxide is given by $C_P = 43.26 + 0.0115T$ where C_P is in kJ/kgmol-K and T is in degrees kelvin.

3-8: Liquid allyl alcohol has a vapor pressure of 53.32 kPa at 80.2 °C and its normal boiling point is 96.6 °C . Calculate its heat of vaporization over the temperature range of 80.2 to 96.6 °C.

3-9: In a refrigeration cycle using HFC-134a, the refrigerant enters the throttling valve as saturated liquid
at 10 barA and 40 °C. The downstream pressure is 1.38 barA. What is the entropy increase (kJ/kg-K) in the expansion.

3-10: Using the following data , calculate the fugacity coefficient for ethane at 522 K and 69 bar pressure.

Temperature K	Pressure bar	Volume m^3/kg	Enthalpy of vapor kJ/kg	Entropy of vapor kJ/kg-K
522	0.7	2.1	1562.1	8.971
522	69.0	0.0197	1520.0	7.633

3-11: The stored energy of a closed system increases by 110 kJ when work is done on the system. in the amount of 200 kJ. Is the heat transferred to or from the system and how much?

Solutions to problems:

l pressure after latches are released and equilibrium is reached

$$+ \frac{100 \times 9.81}{1000 \times 0.002} = 101.3 + 338.3 = 439.6 \ kN/m^2$$

behavior

$$W = \int_{V_1}^{V_2} P_{ext} dV = P_{ext} (V_2 - V_1) = P_m(V_m - V_1)$$

dicate intermediate pressure and volume.

$P_m = 439.6 \ kN/m^2$.

$\quad$ - V_1) $= P_m V_m(1 - P_m/ P_1)$ $\quad$ since $\quad V_1 / V_m = \quad P_m /P_1$
$\quad\quad\quad = nRT (1 - P_m / P_1)$ $\quad$ since $P_m V_m = nRT$
$\quad\quad\quad = 100 \ (8.314) \ (283) \ (1 - 439.6/516.3)$
$\quad\quad\quad = 34955 \ J = 43.955 \ kJ.$

, the gas expands against constant external pressure of 101.3 kN/m^2
ssure is now 439.6 kN/m^2. Then
$T(1 - P_1 /P_m) = 100(8.314)(283) (1 - 101.3/439.6) = 181.076 \ kJ$
$\quad$ Total work done $= 34.955 + 181.076 = 216.03 \ kJ$
given by $W = \int_{V_1}^{V_2} PdV$ $\quad$ (reversible)
gas, at constant temperature $PdV = - VdP = -RT \ \frac{dP}{P}$
$\quad$ Therefore $W = - RT \ln \frac{P_2}{P_1} = - 8.314(283) \ln \frac{101.3}{516.3} = 3.832 \ kJ/g\text{-}mol$
Then work per 100 gmol $= 100 \ (3.832) = 383.2 \ kJ$

3-2:

(a) Carnot efficiency $= \frac{T_H - T_C}{T_H}$
$\quad$ Efficiencies of the two engines are equal. Since $\frac{T_H - T}{T_H} = \frac{T - T_C}{T}$
$\quad$ Therefore, we can get from the equality $T/T_H = T_C/T$ or $T^2 = \sqrt{T_H T_C}$
$\quad T_H = 1111 \ K$ $\quad$ and $\quad T_C = 300 \ K$

Then $T^2 = \sqrt{1111 \times 300} = 577.3$ K

(b) In this case works are equal. Therefore, $W_1 = Q_H - Q_T = W_2 = Q_T - Q_C$

Dividing by Q_T, one obtains $\frac{Q_H}{Q_T} - 1 = 1 - \frac{Q_C}{Q_T}$ (equation 1)

Also $\frac{Q_H}{Q_T} = \frac{T_H}{T}$ and $\frac{Q_C}{Q_T} = \frac{T_C}{T}$.

By substitution of these relations in equation 1 above, we get $\frac{T_H}{T} + \frac{T_C}{T} = 2$

Which yields $T = \frac{T_H + T_C}{2} = \frac{1111 + 300}{2} = 705.$ K

3-3:

Van der Waal's equation is given by $\left(P + \frac{a}{V^2}\right)(V - b) = RT$ for 1 g-mol

This can be expressed as $P = \frac{RT}{V-b} - \frac{a}{V^2}$

Now multiply each side by $\bar{V}$, then $P\bar{V} = \frac{RT\bar{V}}{V-b} - \frac{a\bar{V}}{V^2}$

which can be rearranged as $P\bar{V} = RT\left(1 - b/\bar{V}\right)^{-1} - \frac{a}{\bar{V}}$

Now expand the term in braces using binomial theorem as follows

$$\left(1 - \frac{b}{\bar{V}}\right)^{-1} = 1 + \frac{b}{\bar{V}} + \frac{b^2}{\bar{V}^2} + \frac{b^3}{\bar{V}^3} +$$

Substituting this expansion in the equation and collecting together the terms with the same power of $\bar{V}$, the virial form of Van der Waal's equation results as follows

$$P\bar{V} = 1 + \frac{RT(b-a)}{\bar{V}} + \frac{RTb^2}{\bar{V}^2} + \frac{RTb^3}{\bar{V}^3} +$$

3-4:

Van der Waal's equation $\left(P + \frac{a}{v^2}\right)(v - b) = RT$ wherein specific volume $= v = \bar{v}$

Simplifying the equation, $Pv - Pb + \frac{a}{v} - \frac{ab}{v^2} = RT$

Multiply each term by v^2 and transpose all terms from r.h.s. to l.h.s. and also divide by P.

One obtains the equation : $v^3 - (Pb + RT)v^2 + \frac{a}{P}(v - b) = 0$

$P = 1379$ kPa $R = 8.314 \frac{kPa-m^3}{kgmol-K}$

$a = 757.4$ kPa $b = 0.065$ m^3/kg.mol

Substitution in the equation, there results the following relation

$$v^3 - 2.954v^2 + 0.55(v - 0.065) = 0$$

which is cubical in and must be solved by trial or graphically

The calculations are done as follows.

(Hint: Assume the first trial value obtained from the ideal gas law.)

Assumed value of v	2.9	2.62	2.76
Calculated value of LHS	1.105	-0.87	0.004436 (close to 0)

Therefore $v = 2.76$ m^3/ kgmol

3-5:

Carnot engine is a reversible engine. Therefore, $\frac{Q_H}{T_H} = \frac{Q_C}{T_C}$

Also $W = Q_H - Q_C = 5021$ kJ

T_H 1200 K $T_C = 300$ K

Then $Q_H = Q_C(T_H/T_C) = Q_C (1200/300) = 4Q_C$

Hence $Q_H - Q_C = 4Q_C - Q_C = 5021$ kJ

From which $Q_C = 5021/3 = 1673.7$ kJ

Therefore $Q_H = 4 \times 1673.7 = 6695$ kJ

The engine absorbs 6695 kJ at 1200 K and rejects 1673.7 kJ at 300 K

3-6:

$$T = 1000 \text{ K}$$

At this temperature, assumption of ideal gas behavior is appropriate, and therefore $C_P - C_V = R$

$$C_P = 21.3 + 11.562\text{x}10^{-2}\,T - 3.812\text{x}10^{-5}\,T^2$$
$$= 21.3 + 11.562\text{x}10^{-2}(1000) - 3.812\text{x}10^{-5}(1000)^2 = 98.8 \text{ kJ/kgmol}$$
$$C_V = 98.8 - 8.314 = 90.49 \text{ kJ/kgmol-K}$$
$$C_P/C_V = 98.8/90.49 = 1.092$$

3-7:

Assume CO_2 to behave ideally over the temperature range of the problem

Therefore $C_V = C_P - R = 43.26 + 0.0115T - 8.314$
$$= 34.946 + 0.0115\,T \quad T \text{ in Kelvin.}$$

From first law, $\quad \Delta U = Q = C_V d$
$$T_2 = 700 \text{ K}$$
$$T_1 = 300 \text{ K}$$

Then $\quad Q = \int_{T_1}^{T_2} C_V dT = \int_{300}^{700}(34.946 + 0.0115T)dT$
$$= 34.946(T_2 - T_1) + \frac{0.0115}{2}\left(T_2^2 - T_1^2\right)$$
$$= 34.946(700 - 300) + 0.00575(700^2 - 300^2)$$
$$= 16278.4 \text{ kJ/kgmol}$$

kgmol of CO_2 to be heated $= 25/44 = 0.5682$ kgmol

Heat to be supplied $= 0.5682(16278.4) = 9249.4$ kJ

3-8:

Assume $\Delta H = $ constant and use the equation
$$\ln\frac{P_2}{P_1} = \frac{\Delta H_V}{R}\left(\frac{1}{T_1} - \frac{1}{T_2}\right)$$
$T_1 = 80.2 + 273 = 353.2$ K $\quad T_2 = 96.6 + 273 = 369.6$ K

Then $\quad \ln\frac{101.3}{53.32} = \frac{\Delta H_V}{8.314}\left(\frac{1}{353.2} - \frac{1}{369.6}\right)$

from which $\quad \Delta H_V = 42472 \text{ kJ/kgmol}$

3-9:

The expansion through a throttling valve is isenthalpic. Therefore $h_i = h_o$

From P-H diagram of HFC-134a, (SI system)

For liquid, $h_i = 257$ kJ/kg $\quad S_i = 1.19$ kJ/kg-K

At 1.38 barA, $h_o = 257$ kJ/kg of liquid-vapor mixture

Following constant enthalpy line at 1.38 barA, $S_o = 1.225$ kJ/kg-K

The entropy increase during expansion $= S_o - S_i = 1.225 - 1.19 = 0.035$ kJ/kg-K

3-10:

$$dG_T = RTd\ln \quad \text{or} \quad d\ln f = \frac{dG}{RT}$$

Integration between high pressure P and low pressure at constant temperature gives

$$RT\ln\frac{f}{f^*} = \bar{G} - \bar{G}^* \text{ and since } G = H - T$$

$$\ln\frac{f}{f^*} = \frac{1}{R}\left[\frac{\bar{H}-\bar{H}^*}{T} - (\bar{S}-\bar{S}^*)\right]$$

If the reference state is a low pressure, then $f^* = P^*$ and then

$$\ln\frac{f}{P^*} = \frac{1}{R}\left[\frac{\bar{H}-\bar{H}^*}{T} - (\bar{S}-\bar{S}^*)\right]$$

Assume ethane behaves as an ideal gas at 0.7 barA and 522 K. Then

$P^* = 0.7$ barA , $\quad \bar{H}^* = 1562.1$ kJ/kg, $\quad \bar{S}^* = 8.971$ kJ/kg-K

At 69 barA and 522 K, $\bar{H} = 1520$ kJ/kg and $\bar{S} = 7.633$ kJ/kg-K

Therefore , $\quad \ln\frac{f}{P^*} = \frac{30}{8.314}\left[\frac{1520-1562.1}{522} - (7.633 - 8.971)\right] = 4.533$

Then $\quad \frac{f}{P^*} = e^{4.533} \quad 93.04$ barA and therefore $f = 93.04(0.7) = 65.1$ barA

Fugacity coefficient = $\quad \phi = \frac{f}{P} = 65.1/69 = 0.9435$

3-11:

Energy balance on the closed system gives

$(U + PE + KE)_E - (U + PE + KE)_B = (H + PE + KE)_{II} - (H + PE + KE)_o + Q - W$
$\quad\quad 1 \quad\quad 1 \quad\quad\quad\quad 1 \quad\quad 1 \quad\quad\quad\quad 2 \quad\quad\quad\quad\quad 2$

1 Negligible potential and kinetic energy effects
2 Closed system, therefore no mass transfer across boundaries of the system
Thus energy balance reduces to : $U_E - U_B = Q - W$
Or in difference form , $\quad \Delta U = Q - W$
$\Delta U = + 110$ kJ $\quad W = -200$ kJ since work is done on the system.
Inserting in the final energy balance
$\quad\quad\quad\quad 110 = Q - (-200)$
$\quad$ and $\quad\quad Q = -90$ kJ

Since Q is negative, 90 kJ must be removed from the system.
(Note: The above solution is given at length for the sake of initial clarification. One should be able to write the final energy balance by working out the reasoning mentally, to save considerable time . One should write the energy balance $\quad \Delta U = Q - W \quad$ at once and proceed further with the solution in this example.)

References:
1. Das D. K. and R.K.Prabhudesai, **Chemical Engineering License Review**, Engineering Press, 2nd Ed.,(1996)
2. Smith J. M. and H.C. Van Ness, **Introduction to Chemical Engineering Thermodynamics**, McGraw-Hill, 3rd Ed.,(1975)
3. Thermodynamics, **AIChE modular instruction series vol 1-4.**

CHAPTER 4

MASS TRANSFER

Mass transfer involves movement of molecules under some driving force such as concentration or temperature difference. However, mass transfer discussed here will be primarily dealing with concentration difference. Mass transfer takes place by two mechanisms (1) Molecular diffusion (2) Convective diffusion

Fick's law: This basic relation for mass transfer by molecular diffusion in a binary mixture of A and B is given by
$$J_A = -D_{AB}\frac{dC_A}{dz}$$
Where J_A = Molar flux of component A along axis z and moving at molar average velocity
D_{AB} = Diffusivity of A in the solution containing A and B
C_A = Concentration of A
z = Distance coordinate in the direction of molecular diffusion

Another expression for the same case is
$$J_A = -c_m D_{AB}\frac{dy_A}{dz}$$
where y_A is mol fraction of A and c_m is average molar density of the mixture.

Estimation of binary diffusivities:
For gases, an empirical equation to calculate diffusivity of a component in a binary mixture is as follows

$$D_{AB} = \frac{0.001T^{1.75}\left(1/M_A + 1/M_B\right)^{1/2}}{P\left[(\Sigma v)_A^{1/3} + (\Sigma v)_B^{1/3}\right]^2} \qquad cm^2/s$$

where v is the atomic diffusion volume of a simple molecule and other symbols have usual meaning e.g.
p = pressure. Σv's are obtained by summation of the atomic diffusion volumes.
Theoretical equation for calculation of diffusivity in binary gas mixtures at low-pressures is

$$D_{AB} = \frac{0.001858T^{3/2}\left(1/M_A + 1/M_B\right)^{1/2}}{P\sigma_{AB}^2\Omega_D} \qquad cm^2/s$$

Where T = temperature K, M_A and M_B = molecular weights, P = Pressure, atm,
$\Omega_D = f\left(kT/\in_{AB}\right)$, collision integral, k = Boltzmann's constant,
and $\in_{AB}, \sigma_{AB}$ Lennard-Jones force constants for the binary.
The values of $\in_{AB}$ and σ_{AB} are obtained from values for the pure components by relations
$$\frac{\in_{AB}}{k} = \left(\frac{\in_A}{k}\frac{\in_B}{k}\right)^{1/2} \text{ and } \sigma_{AB} = \tfrac{1}{2}\left(\sigma_A + \sigma_B\right)$$
Values of $\in/k$ and σ are available in literature. If not, they may be estimated by the following

$$\frac{\in}{k} = 0.75T_c \qquad and \qquad \sigma = \tfrac{5}{6}V_c^{1/3}$$

Where T_c = critical temperature , K V_c = critical volume, cm^3/g-mol

σ Lennard-Jones potential parameter, $A°$

For liquids, Wilke and Chang relation is given by

$$D^\circ_{AB} = 7.4 \times 10^{-8} \left[(\phi M_B)^{1/2} \frac{T}{\mu_B V_A^{0.6}} \right]$$

where ϕ= association parameter of solvent B , $\phi = 2.26$ for water as solvent,
1.9 for methanol, 1.5 for ethanol and 1 for unassociated solvents such as benzene and ethyl ether.

$\mu_B =$ viscosity of B, cP

V_A = molar volume of solute A at its normal boiling point, cm^3/g-mol

Diffusivity in a multi-component mixture: The diffusivity of a component in a mixture is expressed by the Stefan-Maxwell equation. A simplified version of the equation is:

$$D_{1-mix} = \frac{1}{\sum\limits_{i=2}^{n} y'_i / D_{1-i}} \quad \text{where} \quad y'_i = y_i / \sum\limits_{j=2}^{n} y_j$$

y'_i = mol fraction of component i , calculated by excluding component 1.

Some relations for mass transfer under non-flow conditions are given below:

Steady state diffusion of A into stagnant gas-film of B (gas):

$$N_{Az} = \frac{D_{AB} P}{RT(z_2 - z_1)} \frac{1}{(p_B)_{lm}} (p_{A1} - p_{A2}) = \frac{c D_{AB}}{(z_2 - z_1)} \ln \frac{x_{B2}}{x_{B1}}$$

Steady state equimolar counter diffusion of A and B: (gas):

$$N_{Az} = \frac{D_{AB}}{z_2 - z_1} (C_{A1} - C_{A2}) = \frac{D_{AB}}{RT(z_2 - z_1)} (p_{A1} - p_{A2})$$

Steady state diffusion of A into stagnant liquid film of B:

$$N_{Az} = \frac{D_{AB} \rho}{M_{av}(z_2 - z_1)} \frac{1}{(x_B)_{lm}} (x_{A1} - x_{A2})$$

Equimolar counter diffusion (liquids) :

$$N_{Az} = \frac{D_{AB}}{(z_2 - z_1)} \left(\frac{\rho}{M} \right)_{av} (x_{A1} - x_{A2})$$

Vaporization of a spherical drop of a liquid:

$$N_{Az} = \frac{\rho D_{AB}}{r_0} \frac{1}{(W_B)_{lm}} (W_{B\delta} - W_{BS})$$

Mass transfer coefficients:

Mass transfer coefficient is defined by the relation

Mass flux = (mass transfer coefficient)(driving force)

For example, for the diffusion of A through non diffusing B ,

$$N_A = k_G (p_{A1} - p_{A2}) = k_y (y_{A1} - y_{A2}) = k_c (C_{A1} - C_{A2}) \qquad \text{for gases}$$

$$N_A = k_x (x_{A1} - x_{A2}) = k_L (C_{A1} - C_{A2}) \qquad \text{for liquids}$$

Units of mass transfer coefficients will depend upon the units chosen for the driving force viz. the concentration difference.

For gases, $k_g = (k_y/P_t) = (k_c/RT)$ since $p_A = y_A P_t$ and $C_A = (p_A/RT)$.

In like manner, for liquids, the relationship is $k_L C = k_x$ since $C_A = C x_A$ where k is a mass transfer coefficient in appropriate units. For example, $k_y = $ moles/(cm^2.s. Δy). Mass transfer coefficients for various situations are correlated in terms of dimensionless numbers as in heat transfer. These are:

$$\text{Sherwood number} = \left(\frac{kD}{D_{AB}}\right) = \text{Sh}$$

$$\text{Reynold's number} = \left(\frac{Du\rho}{\mu}\right) = \text{Re}$$

$$\text{Schmidt number} = \left(\frac{\mu}{\rho D_{AB}}\right) = \text{Sc}$$

$$\text{Grashof number} = \left(\frac{gD^3\rho\Delta\rho}{\mu^2}\right) = \text{Gr}$$

The product of Re and Sc is called Peclet number, Pe and St, Stanton number = Sh/Pe
Depending upon the nature of the mass transfer coefficient used, expressions for the Sherwood number will change. Thus Sherwood number $\text{Sh} = \frac{Fl}{cD_{AB}} = \frac{k_G P_{B,M} RTl}{D_{AB} P_t} = \frac{k_c P_{B,M} l}{P_t D_{AB}}$ etc.

Mass Transfer from a Gas into a Falling Liquid Film:

For Re < 100, $k_{L,av} = 3.41 \frac{D_{AB}}{\delta}$ or $\text{Sh}_{av} = \frac{k_{l,av}\delta}{D_{AB}} = 3.41$ where $\delta = $ film thickness

For Re > 100 $k_{L,av} = \left(\frac{6D_{AB}\Gamma}{\pi\rho\delta L}\right)^{1/2}$ and $\text{Sh}_{av} = \left(\frac{3}{2\pi}\frac{\delta}{L}Re\,Sc\right)^{1/2}$

where $\delta = $ film thickness, $\Gamma = $ mass rate of liquid flow per unit width of film in the x direction., and L = length in the z direction. For a falling film, the film thickness is given by

$$\delta = \left(\frac{3\mu\Gamma}{\rho^2 g}\right)^{1/3} = \left(\frac{3\bar{u_y}\mu}{\rho g}\right)^{1/2}.$$

$k_{L,av}$ is given by : $k_{L,av} = \frac{\bar{u_y}\delta}{L} \ln\frac{C_{Ai}-C_A}{C_{A,i}-\bar{C}_{A,L}}$.

Then flux, $N_{A,av} = k_{L,av}\left(C_{A,i}-\bar{C}_A\right)_M$

where $(C_{A,i}-\bar{C}_{A,L})$ logarithmic average concentration difference over length L.

Mass transfer from spheres:

For stationary fluid : $Sh_o = \frac{k_c D}{D_{AB}} = 2$ where $N_{sho} = $ Sherwood number for stationary fluid.

For convection : $Sh = Sh_o + 0.6\,Re^{0.5}Sc^{1/3}$ [Frossling Equation]

Another equation for Convective mass transfer:

$$\frac{k_c D}{D_{AB}} = 2 + 1.0\,Re^{1/3}Sc^{1/3} = 2 + 1.0\,Pe^{1/3} \quad \text{for Pe} > 1.0$$

Mass transfer coefficients for other geometries such as from flat plates, in laminar flow through tubes, rotating discs are available in the literature.

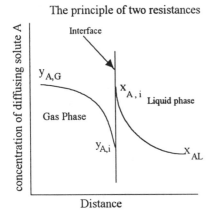
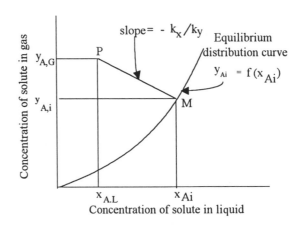

Figure 4-1: Mass transfer between two phases

Turbulent mass transfer:
As in heat and momentum transfer, turbulent mass transfer flux can be written as the sum of two fluxes, one due to turbulence and the other due to laminar boundary as follows

$$\left(N_A\right)_{Total} = -\left(D_{AB} + \varepsilon_D\right)\frac{dC_A}{dy}$$

where ε_D = eddy mass diffusivity.

The film model: $N = -D\left(\frac{dc}{dx}\right)_{x=0} = \frac{D}{\delta}\left(C_0 - C_L\right) = \frac{D}{\delta}\Delta C = k_L(\Delta C)$

The Higbie penetration model:

$$N_{A,av} = 2\left(C_{Ai} - C_o\right)\left(\frac{D_{AB}}{\pi t}\right)^{1/2} \quad \text{and} \quad k_L = 2\left(D_{AB}/t\right)^{1/2}, \quad \text{also } Sh = 1.13\,Re^{0.5}Sc^{0.5}$$

Surface renewal theory of Danckwerts:

$$N_A = \left(C_{Ai} - C_o\right)\sqrt{Ds} \quad \text{where s is fractional surface renewal rate.} \quad k_c = \sqrt{Ds}$$

Toor-Marchellow/Dobbins models are also available.

Boundary layer theory:
The boundary layer theory can be used to explain the turbulent mass transfer. The expressions derived are:

$$k = 0.664 Re^{0.5} Sc^{0.5} \frac{D_{AB}}{L} \quad \text{and} \quad k_L = \frac{D_{AB}}{L}(0.0365)(0.302) Re^{1.08} Sc^{7/15}$$

Analogy correlations (heat transfer and momentum transfer) and J factor methods are also applied.

Mass transfer coefficients correlations for turbulent flow:

Motion of fluid	Range of condition	Equation
Inside tube (circular)	Re = 4000 - 60000 Sc = 0.6 - 3000	$j_D = 0.023 Re^{-0.17}$ $Sh = 0.023\, Re^{0.83} Sc^{1/3}$
Flow parallel to flat plate (unconfined)	Re = 10000-400000 Sc > 100	$j_D = 0.0149 Re^{-0.12}$ $Sh = 0.0149 Re^{0.88} Sc^{1/}$
Liquid film in wetted-wall tower,	$4\Gamma/\mu = 0 - 1200$	See equations for laminar flow.
Transfer between liquid and gas	$4\Gamma/\mu = 1300 - 8300$	$Sh = 1.76 \times 10^{-5} \left(\frac{4\Gamma}{\mu}\right)^{1.506} Sc^{0.5}$
	Re = 4000-30000	$\frac{k_c d}{D_{AB}} \frac{p_{BM}}{P} = 0.023\, Re^{0.83} Sc^{0.44}$
		or $\frac{k_c d}{D_{AB}} \frac{p_{BM}}{P} = 0.0328\, Re^{1.1} Sc^{1/3}$
Perpendicular to single cylinders	Re = 400 - 25000	Gases: $Sh = 0.28 Re^{0.5} Sc^{0.44}$ Liquids: $Sh = 0.281 Re^{0.6} Sc^{1/3}$
Flow past single spheres:	Re = 2000-17000	$Sh = 0.347 Re^{0.62} Sc^{1/3}$

Lewis and Whitman assumed that in mass transfer between two fluid phases, resistances to mass transfer occur only in the fluids and there is no resistance to solute transfer across the interface between the two phases. This is called two-resistance principle. (see the Figure Mass 4-1).

At the interface , $y_{A,i}$ and $x_{A,i}$ are in equilibrium. If k_x and k_y are local mass transfer coefficients, $N_A = k_y(y_{AG} - y_{Ai}) = k_x(x_{Ai} - x_{AL})$ where $(y_{AG} - y_{Al})$ and $(x_{Ai} - x_{Al})$ are the driving forces for mass transfer in the gas and the liquid phases. A rearrangement gives the relation: $\frac{(y_{Ag} - y_{Ai})}{(x_{AL} - x_{Al})} = -\frac{k_x}{k_y}$

Thus if the mass transfer coefficients are known, the interfacial concentrations and the flux N_A can be determined graphically or analytically by solving the above equation in conjunction with the equilibrium curve.

Overall mass transfer coefficients:
As in heat transfer, in mass transfer also, one can write overall mass transfer coefficients in terms of the individual mass transfer coefficients in terms of the bulk concentrations. With liquid phase as basis, it is possible to derive the following relation

$$\frac{1}{K_x} = \frac{1}{mk_y} + \frac{1}{k_x}$$

If gas phase is chosen as basis to derive the overall mass transfer coefficients, the relation is

$$\frac{1}{K_y} = \frac{1}{k_y} + \frac{m}{k_x}$$

where m is the slope of the equilibrium curve assumed to be a straight line. It should be noted that if the gas phase resistance controls, $\frac{1}{K_y} \approx \frac{1}{k_y}$

and if the liquid phase resistance controls, $\frac{1}{K_x} \approx \frac{1}{k_x}$.

These relations imply that $\quad y_{Ag} - y*_{Ai} \approx y_{Ag} - y_{Ai} \qquad$ for gas phase controlling

and $\qquad x_A^* - x_{AL} \approx x_{Ai} - x_A \qquad$ for liquid phase controlling

where $*$ denotes equilibrium concentrations and y_{AG} x_{Al} are bulk phase concentrations. In practice average mass transfer coefficients are used and these are correlated by dimensionless numbers.

For general case where (1) mass transfer rates are high (2) diffusion of more than one substance is involved or (3) a situation in which equimolar counter-diffusion is not involved F_G and F_L should be used. Thus mass-transfer flux is given by

$$N_A = \frac{N_A}{\Sigma N} F_G \ln \frac{N_A' \Sigma N - y_{Ai}}{N_A' \Sigma N - y_{Ag}} = \frac{N_A}{\Sigma N} F_L \ln \frac{N_A' \Sigma N - x_{AL}}{N_A' \Sigma N - x_{Ai}}$$

where F_G and F_L are generalized gas- and liquid- phase mass transfer coefficients for component A. One can also define generalized overall mass transfer coefficients F_{OG} and F_{OL}.

The following relations can be derived for two simple cases.
1. **Diffusion of one component :**

$$e^{N_A'F_{OG}} = e^{N_A'F_G} + m' \frac{1 - x_{A,L}}{1 - y_{A,G}} (1 - e^{-N_A'F_L}) \quad \text{and} \quad e^{-N_A'F_{OL}} = \frac{1}{m''} \frac{1 - y_{AG}}{1 - x_{Al}} \left(1 - e^{N_A'F_G}\right) + e^{-N_A'F}$$

2. **Equimolar counter-diffusion :**

$$\frac{1}{F_{OG}} = \frac{1}{F_G} + \frac{m'}{F_L} \qquad \text{and} \qquad \frac{1}{F_{OL}} = \frac{1}{m''F_G} + \frac{1}{F_G}$$

where m' , m" are slopes of chords of the equilibrium curve, dimensionless

Mass transfer in packed beds:
Mass transfer in packed beds takes place between two flowing phases. Usual approach is to determine the product of the mass transfer coefficient and the interfacial area per unit volume of the packed bed. If a is the interfacial area per unit volume of packed bed, $\mathbf{a} = \frac{A_i}{V_t}$ where A = total interfacial area in a packed tower of volume V_t. Then molar flow rate W_A of the diffusing species A is given by

$$W_A = k_Z a V_t (Z_{AS} - Z_A) \quad \text{where} \quad k_Z \text{ can be either } k_x a \text{ or } k_y a.$$

However, mass transfer coefficients in packed towers are correlated in terms of a height of a transfer unit, H. The height of transfer unit is defined as follows

$$H_G = \frac{G}{k_y a}$$ where G = moles of gas/(unit time. unit cross section of empty tower).

Similarly,

$$H_L = \frac{L}{k_x a}$$ where L = mols of liquid/ (unit time. unit cross section of empty tower)

H can also be defined to include the overall mass transfer coefficient as follows:

$$H_{OG} = \frac{G}{K_y a}$$ (Overall height of transfer unit based on gas phase)

$$H_{OL} = \frac{L}{K_x a}$$ (Overall height of transfer unit based on liquid phase)

Analytical expressions for H_{OG} and H_{OL} :

$$H_{OG} = \frac{G}{K_y a (1-y)_{lm}} = \frac{G}{K_G a P_t (1-y)_{lm}}$$ where $$(1-y)_{lm} = \frac{(1-y^*) - (1-y)}{\ln \frac{(1-y^*)}{(1-y)}}$$

$$H_{OL} = \frac{L}{K_x a (1-x)_{lm}} = \frac{L}{K_L a C (1-x)_{lm}} = \frac{L M_L}{K_L a \rho_L (1-x)_{lm}}$$ where $$(1-x)_{lm} = \frac{(1-x) - (1-x^*)}{\ln \frac{1-x}{1-x^*}}$$

M_L = Molecular weight of liquid solvent.

The overall mass transfer coefficients $K_G a$ and $K_y a$ can be calculated from the individual gas and liquid phase mass transfer coefficients and equilibrium relationship. Table mass 1-1 will be useful to identify correct units to be used in a particular case.

Table Mass 1-1: Units of mass transfer coefficients

(a) $\frac{1}{K_G a} = \frac{1}{k_G a} + \frac{m_c}{k_L a}$ Here m_c is given by the equilibrium relationship $p^* = m_C C$

where p^* is in atm., and C in lb-mol/ft^3.

(b) $\frac{1}{K_G a} = \frac{1}{k_G a} + \frac{1}{m'_c k_L a}$ In this equation, m'$_c$ is given by $p^* = C/m'_C$

(c) $\frac{1}{K_G a} = \frac{1}{k_G a} + \frac{m p_t}{k_L a \rho_m}$ where m is given by $y^* = mx$

and $\rho_m = \frac{density \ of \ solution, \ lb/ft^3}{molecular \ weight \ of \ solution}$

(d) $\frac{1}{K_G a} = \frac{1}{k_G a} + \frac{m_x}{k_L a \rho_m}$ where m_x is given by $\rho_M p^* = m_x x$

(e) $\frac{1}{K_y a} = \frac{1}{k_y a} + \frac{m}{k_x a}$ where m is given by $y^* = mx$

(f) $K_G a = K_y a / p_t$

(g) $k_G a = k_y a / p_t$

Liquid phase mass transfer coefficients may also be expressed in different units as follows:

(a) $\dfrac{1}{K_L a} = \dfrac{1}{k_L a} + \dfrac{m_c'}{k_G a}$ (b) $\dfrac{1}{K_L a} = \dfrac{1}{k_L a} + \dfrac{1}{m_c k_G a}$ (c) $\dfrac{1}{K_L a} = \dfrac{1}{k_L a} + \dfrac{\rho m}{mk_G ap_t}$

(d) $\dfrac{1}{K_L a} = \dfrac{1}{k_l a} + \dfrac{\rho m}{m_x k_G a}$ (e) $K_L a = K_x a/C$ (f) $k_L a = k_x a/C$

By analogy with H_{OL} and H_{OG}, the gas film and liquid film transfer units can also be defined as follows:

(a) $H_{tG} = \dfrac{G}{k_y a \left(1-y\right)_{lm}} = \dfrac{G}{k_y a}$ (b) $H_{tG} = \dfrac{G}{k_G a P_t \left(1-y\right) lm} \approx \dfrac{G}{k_G a P_t}$

(c) $H_{tL} = \dfrac{L}{k_x a(1-x)_{lm}} \approx \dfrac{L}{k_x a}$ (d) $H_{tL} = \dfrac{L}{k_L a C(1-x)_{lm}} \approx \dfrac{L}{k_L a C}$

Diffusion in solids:
Fick's law can be applied for diffusion in solids under the following conditions; (1) concentration gradient is independent of time (2) diffusivity is constant and independent of concentration (3) and there is no bulk flow. Then , the flux is given by $N_A = -D_A \dfrac{dC}{dZ}$ where D_A is the diffusivity of A through the solid. For constant D_A, the following relations are obtained:

Diffusion through a flat slab: $N_A = \dfrac{D_A\left(C_{A1}-C_{A2}\right)}{Z}$

Other solid shapes: $w = N_A S_{av}\left(C_{A1} - C_{A2}\right)/Z$

where s_{av} = average cross section for diffusion.

Radial diffusion through a solid cylinder of inner and outer radii r_1 and r_2 and of length L

$$S_{av} = \dfrac{2\pi L\left(r_2 - r_1\right)}{\ln\left(r_2/r_1\right)} \quad \text{and} \quad Z = r_2 - r_1$$

Radial diffusion through a spherical shell: with inner and outer radii r_1 and r_2
$$S_{av} = 4\pi r_1 r_2 \quad \text{and} \quad z = r_2 - r_1$$

Diffusion in porous solids:
Effective diffusivities are to be used in place of diffusivity in Fick's law equation.

Diffusion through polymers: The diffusion is expressed in terms of permeability P as follows:
$$V_A = \dfrac{D_A S_A\left(\bar{p}_{a1} - \bar{p}_{a2}\right)}{Z}$$

where V_A = diffusional flux, cm^3 gas (STP) / (cm^2.s) D_A = diffusivity of A, cm^2/s
S_A = solubility coefficient or henry's law constant, cm^3 gas (STP)/(cm^3 solid). cmHg
Z = thickness of polymeric membrane, cm
$P = D_A S_A$ where P = permeability, cm^3 gas (STP) / (cm^2.s)(cmHg/cm)

Example Problems:

Mass 4-1: Calculate the rate of diffusion NaCl across a film of water (non-diffusing) solution 1.5 mm thick at 18 °C when the concentrations on opposite sides of the film are 24 and 4 wt % NaCl respectively. The diffusivity of NaCl is 1.3×10^{-5} cm^2/s at 18 °C. Densities of 24 and 4 wt % NaCl at 18 °C are ~~1181~~ 1081 and 1027 kg.m^3.

Mass 4-2: A rectangular tank 0.1 m^3 in cross section and 15 cm depth is filled with ethyl alcohol solution in water at 10 °C to within 2 cm of the top. A fan is blowing air at 10 °C and 1 bar over the tank. Alcohol concentration at the interface is 0.55 mol fraction. The diffusivity of ethyl alcohol is 0.039 m^2/h. How much alcohol will be lost in one day by diffusion? Assume that alcohol level is maintained constant, and local atm. pressure is 1 barA.

Mass 4-3: Air is flowing in a tube whose inside surface is wetted with water. The temperature is 21 °C. At a point where y_A, the bulk concentration (mol fraction) of H$_2$O is 0.0015. Calculate the flux of water evaporated from the pipe wall. Additional data are as follows: ID of tube = 10.2 cm, air velocity = 15.3 m/s, pressure = 1 barA., D_{AB} = 0.2 cm^2/s, μ of air at 21 °C = 0.018 mPa s, density of air = 1.2 kg/m^3. Sh = $0.023 Re^{0.83} Sc^{0.33}$, $N_A = K_G(P_{A1} - P_{A2})$, $K_G = K_C/(RT)$

Mass 4-4: Hydrogen gas at 15.2 barA and 30 °C is transported through a steel pipe with ID and OD of 58.5 and 89 mm, respectively to a reaction system. Henry's law constant for the solubility of hydrogen in steel is reported to be 1.67 bar.m^3/kmol. The diffusion coefficient of hydrogen in steel is 0.3×10^{-12} m^2/s.
Calculate the mass flux of loss of hydrogen by diffusion per hour per 100 meter length of pipe in units of kg/h.

Solutions to problems:
Mass 4-1:

$Z = .0015$ m, $M_A = 58.5$, $M_B = 18.02$, $D_{AB} = 1.3 \times 10^{-5}$ cm^2/s $= 1.3 \times 10^{-9}$ m^2/s
density of 24% solution is 1081 kg/m^3. Therefore,

$x_{A1} = \dfrac{0.24/58.5}{0.24/58.5 + 0.76/18.02} = \dfrac{0.0041}{\underset{0.0463}{0.054}} = 0.0886$ $x_{B1} = 1 - x_{A1} = 1.0.0886 = 0.9114$

$M = \dfrac{1}{0.0463} = 21.$ kg/kmol $\dfrac{\rho}{M} = \dfrac{1081}{21.6} = 50.0$ kmol/m^3.

Similarly for 4 % solution,

$x_{A2} = \dfrac{0.04/58.5}{0.04/58.5 + 0.96/18.02} = \dfrac{0.000684}{0.054} = 0.012$ $x_{B2} = 1 - 0.0127 = 0.9873$

$M = \dfrac{1}{0.054} = 18.5$ $\dfrac{\rho}{M} = \dfrac{1027}{18.52} = 55.4$ kmol/m^3

$\left(\dfrac{\rho}{M}\right)_{AV} = \dfrac{50.05 + 55.45}{2} = 52.7 =$ kmol/m^2/s $x_{BM} = \dfrac{0.9873 - 0.9114}{\ln \frac{0.9873}{0.9114}} = 0.948$

$N_A = \dfrac{D_{AB}}{Z x_{BM}}\left(\dfrac{\rho}{M}\right)_{AV}\left(x_{A1} - x_{A2}\right) = \dfrac{1.3 \times 10^{-9}}{0.0015 \times 0.9488} \times (52.75)(0.0886 - 0.0127)$
$= 3.657 \times 10^{-6}$ kmol/m^2/s

Mass 4-2 :

Assume air contains no alcohol. Therefore , $p_{A2} = 0$ bar

The open space of 2 cm is filled with stagnant air. Thus the thickness of film is 2 cm = 0.02 m

The partial pressure of alcohol at the interface = 0.55 bar = p_{A1}

Therefore $p_{B1} = 0.45$ bar   and $p_{B2} = 1$ bar.

$$(p_B)_{lm} = \frac{p_{B2} - p_{B1}}{\ln(p_{B2}/p_{B1})} = \frac{1.0 - 0.45}{\ln(1/0.45)} = 0.689 \text{ bar} \qquad T = 273 + 10 = 283 \text{ K}$$

$$N_A = \frac{D_{AB} \times P}{RT} \times \frac{(p_{A1} - p_{A2}) \times A}{L \times (p_B)_{lm}} = \frac{0.039 \times 1 \times 0.1(0.55 - 0)}{0.08314 \times 283 \times 0.02 \times 0.689} = 6.62 \times 10^{-3} \quad \text{kg-mol/h}$$

Alcohol lost per day = $6.62 \times 10^{-3} \times 24 = 0.16$ kgmol/day

Mass 4-3:

$$Re = \frac{du\rho}{\mu} = \frac{0.102(15.3)(1.2)}{0.018 \times 10^{-3}} = 10404 \qquad D_{AB} = 0.2 \text{ cm}^2/\text{s} = 0.072 \text{ m}^2/\text{h}.$$

$$Sc = \frac{\mu}{\rho D_{AB}} = \frac{0.018 \times 10^{-3}(3600)}{1.2(0.072)} = 0.75$$

$$Sh = 0.023 \, Re^{0.83} Sc^{1/3} = 0.023(104040)^{0.83}(0.75)^{1/3} = 305$$

or $Sh = \dfrac{k_c d}{D_{AB}} = 305$

Therefore $k_c = \dfrac{0.072}{0.102} \times 305 = 215.3$ m/h

Vapor pressure of water at 21 °C = 0.025 bar

$p_{B1} = 1 - 0.025 = 0.975$ bar $p_{B2} = 1 - 0.0015 = 0.9985$ bar

$$p_{BM} = \frac{p_{B1} - p_{B2}}{\ln(p_{B1}/p_{B2})} = \frac{0.9985 - 0.975}{\ln(0.9985/0.975)} = 0.986 \text{ bar}$$

$$k_G = \frac{k_c}{RT} = \frac{215.3}{0.08314 \times 294} = 8.81 \, kgmol/(h.m^2 \cdot bar) \quad \text{kg-mol/(h.m}^2\text{.bar)}$$

$$N_A = k_G(p_{A1} - p_{A2}) = 8.81 \left[\frac{kgmol}{h.m^2.bar} \right] \times (0.025 - 0.0015) \text{ (bar)}$$
$$= 0.207 \text{ kgmol/h.m}^2$$

Mass 4-4:

Assume Henry's law applies at 15 barA. pressure. Henry's law: $p = Hc$

Then concentration at the inner surface of pipe = $p/H_C = 15/1.67 = 8.982$ kmol/m³

The concentration at the outer surface of the pipe can be taken as zero.

Since hydrogen is diffusing through radially, the effective average area for diffusion must be estimated. For a cylinder, the average diffusional area is given by

$$S_{av} = \frac{2\pi L(r_2 - r_1)}{\ln(r_2/r_1)}$$

where r_2 and r_1 are outer and inner radii respectively.

$r_2 = 89 \times 10^{-3}$ m   and $r_1 = 58.5 \times 10^{-3}$ m

Therefore $S_{av} = \dfrac{2\pi(100)(89 \times 10^{-3} - 58 \times 10^{-3})/2}{\ln(89 \times 10^{-3}/58.5 \times 10^{-3})} = 22.835$ m² /100 m length

$$Z = \tfrac{1}{2}(89 - 58.5) = 15.25 \text{ mm} = 15.25 \times 10^{-3} \text{ m}$$

Molar flux (assuming diffusivity to be constant) is given by

$$N_A = \frac{0.3 \times 10^{-12} \times 22.835(8.982-0)}{15.25 \times 10^{-3}} = 4.035 \times 10^{-9} \text{ kgmol/s}$$

Mass flux $= 4.035 \times 2 \times 10^{-9} = 8.07 \times 10^{-9}$ kg/s H_2 per 100 m length.
Therefore, mass flux per hour $= 8.07 \times 10^{-9} \times 3600$
$= 2.905 \times 10^{-5}$ kg/h. per 100 m length.

References:

1. Das D. K., and R. K. Prabhudesai, Chemical Engineering License Review,
        Engineering Press,2nd Ed.,(1996)
2. Sherwood T.K., R.L.Pigford and C.R. Wilke, Mass Transfer, McGraw-Hill,(1975)
3. Treybal R. E.,Mass Transfer Operations, McGraw-Hill, 3rd Ed.,(1980)
4. AIChE Modular Instruction Series, Mass Transfer, vol 1-4
5. Perry R. H., and D. W. Green, Perry's Handbook for Chem. Engrs.,
        McGraw-Hill, 50th Ed.,(1984)
6. Greencorn R. A. and D. P. Kessler, Transfer Operations, McGraw-Hill , New York,(1972)

CHAPTER 5

CHEMICAL KINETICS

Heat of chemical reaction = Change in enthalpy of the reaction system at constant pressure. This is a function of the nature of reactants and products involved and also their physical states.
Standard heat of reaction = change of enthalpy when the reaction takes place at 1 atm. in such a manner that it starts and ends with all involved materials at in their normal states of aggregation at a constant temperature of 25 ^{0}C.

Heat of formation of a chemical compound is the change in enthalpy when elements combine to form the compound which is the only reaction product. Standard heat of formation of a compound, ΔH_f is the heat of reaction at 25 ^{0}C and 1 atm.

Standard heat of combustion, ΔH_c = the enthalpy change resulting from the combustion of a substance in its normal state at 25^{0}C and atmospheric pressure, with the combustion beginning and ending at 25^{0}C. Generally gaseous CO_2 and liquid water are the combustion products.

Standard heat of reaction, ΔH_R can be calculated from the heats of formation of the reactants and the products by the relation

$$\Delta H_R = \Sigma \Delta H_{f(products)} - \Sigma \Delta H_{f(reac\tan ts)} \qquad \text{at } 25 \,^0\text{C, 1 atm.}$$

It can also be calculated from the heats of combustion of the reactants and products by the relation

$$\Delta H_R = \Sigma \Delta H_{c(reac\tan ts)} - \Sigma \Delta H_{c(products)} \qquad \text{at } 25 \,^0\text{C, 1 atm.}$$

Heat of reaction at any temperature

$$\Delta H_T = \Delta H_{T_0} + \int_{T_0}^{T} \Delta C_P dT \quad \text{where} \quad \Delta C_P = \Sigma(n_i C_{pi})_{products} - \Sigma(n_i C_{pi})_{reac\tan ts}$$

If $C_P^0 = a + \beta T + \gamma T^2 + \cdots$ for each component taking part in the reaction, and if the reaction is of the type: $\quad n_aA + n_bB + \cdots \rightarrow n_cC + n_dD + \cdots$

Then $\quad \Delta H_T = I_H + \Delta a T + \Delta \beta T^2/2 + \Delta \gamma T^3/3 + \cdots \quad$ where I_H is a constant of integration.
Effect of pressure on heat of reaction is generally negligible.

Chemical equilibrium:

Standard free energy change for a reaction is given by $\quad \Delta G^0 = -RT \ln K$
where K = Equilibrium constant, T = absolute temperature, and R = Gas law constant.

For reaction of the type $aA + bB \rightarrow cC + dD$, the equilibrium constant $K = \dfrac{a_C^c \, a_D^d}{a_A^a \, a_B^b}$

where a's are activities of the components. $a_i = \dfrac{f_i}{f_i^0}$.

If the standard state is unit fugacity i.e. $f_i^o = 1$, $\qquad K = \dfrac{f_C^c f_D^d}{f_A^a f_B^b}$

For ideal gas behavior, $K_P = \dfrac{p_C^c p_D^d}{p_A^a p_B^b} \qquad$ where p's are partial pressures.

In terms of mol fractions, $K_P = \dfrac{(y_c P_t)^c (y_d P_t)^d}{(y_a P_t)^a (y_b P_t)^b} = K_y P_t^{[(c+d)-(a+b)]}$

If $P_t = 1$ atm, $K_P = K_y$

Van't Hoff gives $\dfrac{d\ln K}{dT} = \dfrac{\Delta H_T^0}{RT^2}$.

If ΔH_T^0 is independent of temperature, integration of Vant Hoff's equation gives the following relation

$$\ln \frac{K_2}{K_1} = \frac{-\Delta H_T^0}{R}\left(\frac{1}{T_2} - \frac{1}{T_1}\right)$$

If ΔH_T^0 varies with temperature, $\ln K = \dfrac{-I_H}{RT} + \dfrac{\Delta a}{R}\ln T + \dfrac{\Delta \beta}{R}(\frac{1}{2})T + \dfrac{\Delta \gamma}{R}(\frac{1}{6})T^2 \cdots + I$

$$\Delta G_T^0 = I_H + I_G T - \Delta a T \ln T - \tfrac{1}{2}\Delta \beta T^2 - \tfrac{1}{6}\Delta \gamma T \quad \text{where } I_G = -(IR)$$

(a) $\Delta G_T^0 < 0$, Spontaneous reaction. (b) $\Delta G_T^0 = 0$, then equilibrium (c) $\Delta G_T^0 > 0$, no reaction

Rates of reaction (Homogeneous Reactions)

Based on reactant, rate of reaction $= -\ r = -\dfrac{1}{V}\dfrac{dn}{dt}$

Based on product, rate of reaction $= r = \dfrac{1}{V}\dfrac{dn}{dt}$

For general reaction, aA + bB $\rightarrow$ cC + dD, $\quad -\dfrac{1}{a}\dfrac{dn_A}{dt} = -\dfrac{1}{b}\dfrac{dn_B}{dt} = \dfrac{1}{c}\dfrac{dn_c}{dt} = \dfrac{1}{d}\dfrac{dn_D}{dt}$

if ξ is extent of reaction, $d\xi = -\dfrac{dn_A}{a} = -\dfrac{dn_B}{b} = \dfrac{dn_C}{c} = \dfrac{dn_D}{d} = \dfrac{dn_i}{v_i}$

where v_i is the stoichimetric coefficient of species i. Then $r_i = \dfrac{v_i}{V}\dfrac{d\xi}{dt}$ also $\dfrac{r_i}{v_i} = \dfrac{1}{v_i}\dfrac{dC_i}{dt}$

Conversion X is related to extent of reaction by $\quad X_A = \dfrac{a\xi}{\left(n_A\right)_o}$

Rate in terms of X is given by $\quad -\ r = \dfrac{1}{V}\dfrac{dX_A}{dt}$, $\quad$ X = fraction converted, x = moles converted.

Rate equations in terms of concentration

Reaction	order	Rate equation	solution of integral	Half-life
A $\rightarrow$ P	0	$-\dfrac{dC_A}{dt} = k_0$	$C_{A0} - C_A = k_0\,t$	$t_{1/2} = \dfrac{C_{A0}}{2k_0}$
A $\rightarrow$ P	1	$-\dfrac{dC_A}{dt} = k_1 C_A$	$\ln \dfrac{C_A}{C_{A0}} = -k_1 t$	$t_{1/2} = \dfrac{1}{k_1}\ln 2$
2A $\rightarrow$ P	2	$-\dfrac{dC_A}{dt} = k_2 C_A^2$	$\dfrac{1}{C_A} - \dfrac{1}{\left(C_A\right)_0} = k_2 t$	$t_{1/2} = \dfrac{1}{k_2 C_{A0}}$

Reversible reactions

Reaction $A \rightleftarrows B$ is first order in both directions. The rate equation is

$$-\frac{dC_A}{dt} = \frac{dC_A}{dt} = k_1 C_A - k_2 C_B$$

The solution of this equation in terms of equilibrium concentration is given by

$$\frac{C_A - C_{Ae}}{C_{Ao} - C_{Ae}} = e^{-k_R t}$$

Where $\quad k_R = \dfrac{k_1(K+1)}{K},\qquad C_{Ae} = \dfrac{C_{Bo}+C_{Ao}}{K+1},\quad$ and $\quad K = \dfrac{k_1}{k_2}..$

Analysis of complex rate equations

Parallel reactions $\quad A \xrightarrow{k_1} B \qquad A \xrightarrow{k_2} C$

Rate equations: **(1)** $-\dfrac{dC_A}{dt} = \left(k_1+k_2\right)C_A$ **(2)** $\dfrac{dC_B}{dt}=k_1 C_A$ and **(3)** $\dfrac{dC_C}{dt}=k_2 C_A$

With the condition that at t = 0, $C_A = C_{A0}$ and $C_B = C_C = 0$,

$$x_B = \frac{C_B}{C_{A0}} = \frac{k_1}{k_1+k_2} = \left(1-\frac{C_A}{C_{AO}}\right) = \frac{k_1}{k_1+k_2}x_t \quad \text{and} \quad x_C = \frac{C_C}{C_{A0}} = \frac{k_2}{k_1+k_2}\left(1-\frac{C_A}{C_{A0}}\right)=\frac{k_2}{k_1+k_2}x_t$$

where x_t is the total conversion of A to both B and C. The selectivity of B is: $S_B = \dfrac{x_B}{x_C} = \dfrac{k_1}{k_2}$

Consecutive reactions: $\qquad A \xrightarrow{k_1} B \xrightarrow{k_2} C$

Initial condition: at t = 0, $C_A = C_{A0}$, $C_B = C_C = 0$

The rate equations are (1) $\dfrac{dC_A}{dt}=-k_1 C_A$ (2) $\dfrac{dC_B}{dt}=k_1 C_A - k_2 C_B$ (3) $\dfrac{dC_D}{dt}=k_2 C_B$

Solution: $C_A = C_{A0}e^{-k_1 t}, C_B = C_{A0}k_1\left\{\dfrac{e^{-k_1 t}}{k_2-k_1}-\dfrac{e^{-k_2 t}}{k_1-k_2}\right\}$ and $C_D = C_{A0}\left\{1+\dfrac{k_2 e^{-k_1 t}}{k_1-k_2}+\dfrac{k_1 e^{-k_2 t}}{k_1-k_2}\right\}$

Maximum concentration of B occurs at $t_{max}= \dfrac{\ln(k_2/k_1)}{k_2-k_1}$ and $\dfrac{C_{B\,max}}{C_{A0}}=\left(\dfrac{k_1}{k_2}\right)^{k_2/(k_2-k_1)}$

Homogeneous catalyzed reactions

Reaction $\qquad A \xrightarrow{k_1} P \qquad\qquad A+C \xrightarrow{k_2} P + C \qquad$ where $\quad$ C = catalyst

Reaction rates $\quad$ (1) $\quad -(\dfrac{dC_A}{dt})_1 = k_1 C_A$ (2) $\quad -\left(\dfrac{dC_A}{dt}\right)_2 = k_2 C_A C_C$

Solution $\qquad -\ln\dfrac{C_A}{C_{Ao}} = -\ln\left(1-X_A\right)=\left(k_1+k_2 C_C\right)t$

Autocatalytic reactions

Reaction $\quad A + R \rightarrow R + R \qquad$ Rate equation $\quad -\dfrac{dC_A}{dt}=kC_A C_R$

Solution $\qquad \ln\dfrac{C_R/C_{Ro}}{C_A/C_{Ao}} = \left(C_{Ao}+C_{Ro}\right)kt = C_o kt$

Rate expressions in terms of conversion X

Irreversible Reactions

First order: Reaction $\quad$ A→P $\quad$ Rate expression $\quad \dfrac{dX_A}{dt}=k\left(1-X_A\right)$

Solution : $-\ln\left(1-X_A\right) = kt$ at t=0, $X_A = 0$

Second Order: Reaction ; A + B $\rightarrow$ P

Rate expression: $C_{Ao}\dfrac{dX_A}{dt} = kC_{Ao}^2\left(1-X_A\right)\left(M-X_A\right)$ $M = C_{Bo}/C_{Ao}$

when $M \neq 1$, solution is $\ln\dfrac{M-X_A}{M\left(1-X_A\right)} = C_{Ao}(M-1)kt$

when $M = 1$, , rate expression $\dfrac{dX_A}{dt} = kC_{Ao}\left(1-X_A\right)^2$

Solution with B.C. t = 0, $X_A = 0$, $\dfrac{1}{C_{Ao}}\dfrac{X_A}{1-X_A} = kt$

nth order: Rate expression: $C_{A0} = \dfrac{dX_A}{dt} = kC_{A0}^n\left(1-X_A\right)^n$

B.C. t = 0, $X_A = 0$;   Solution: $\left(1-X_A\right)^{1-n} - 1 = (n-1)C_{A0}^{n-1}kt$

Zero order   Rate expression $\dfrac{dX_A}{dt} = k$ B.C $t = 0$, $X_A = 0$;   solution : $C_{A0}X_A = kt$

Reactions in parallel: $A \overset{k_1}{\rightarrow} B$ $A \overset{k_2}{\rightarrow} C$ B.C. at t = 0, $C_A = C_{A0}$, $C_B = C_{B0}$ and $C_C = C_{C0}$

Solution : $-\ln(1 - X_A) = (k_1 + k_2)\,t$

Reversible Reactions:

First order reversible: $A \overset{k_2}{\underset{k_1}{\rightleftarrows}} B$ Rate expression : $\dfrac{dX_A}{dt} = \dfrac{k_1(M+1)}{M + X_{Ae}}\left(X_{Ae} - X_A\right)$ $M = C_{B0}/C_{A0}$

Solution $-\ln\left(1 - \dfrac{X_A}{X_{Ae}}\right) = \dfrac{M+1}{M + X_{Ae}}k_1 t$ where X_{Ae} is equilibrium conversion.

Second order reversible: $A + B \overset{k_2}{\underset{k_1}{\rightleftarrows}} R + S$

Rate equation : $\dfrac{dX_A}{dt} - k_1 C_{Ao}^2\left(1-X_A\right)^2 - \dfrac{k_2}{k_1}X_A^2$

Solution : $\ln\dfrac{X_{Ae} - \left(2X_{Ae} - 1\right)X_A}{X_{Ae} - X_A} = 2k_1 C_{Ao}\left(\dfrac{1}{X_{Ae}} - 1\right)t$

Michaelis-Menton equation: $-r_A = -\dfrac{dC_A}{dt} = k_1\dfrac{C_A}{1 + k_2 C_A}$

Effect of temperature on rate of reaction

Van't Hoff relation $\dfrac{d\ln k}{dt} = \dfrac{E}{RT^2}$ Arrhenius relation : $k = ae^{-E/RT}$

Reactor design for homogeneous reactions
Batch reactor: Material balance for a component results in $(-r_A)V = N_{AO}\dfrac{dX_A}{d\theta}$ $\theta =$ time

Solution : $\quad \theta = N_{AO} \int_0^{X_A} \dfrac{dX_A}{(-r_A)V}$

When density is constant, $\quad \theta = C_{AO} \int_0^{X_A} \dfrac{dX_A}{-r_A} = -\int_{C_{AO}}^{C_A} \dfrac{dC_A}{-r_A}$

For reactions in which volume changes linearly with conversion, $\quad \theta = C_{AO} \int_0^{X_A} \dfrac{dX_A}{(-r_A)(1 + \varepsilon_A X_A)}$

where ε_A = fractional volume change between no conversion and complete conversion

Flow reactors

For flow reactors, space time and space velocity are performance measures. They are defined as follows

Space-time:

$\tau = \dfrac{1}{S} = $ Time required to process one reactor volume of feed measured at specified conditions
$\qquad = $ [time]

Space velocity: The space velocity is related to space-time in the following manner

$S = \dfrac{1}{\tau} = $ Number of reactor volumes of feed at specified conditions that can be treated in unit time.
$\qquad = $ (time)$^{-1}$

For the conditions of the feed, $\quad \tau = \dfrac{1}{S} = \dfrac{C_{AO}V}{F_{AO}} = \dfrac{(moles\ A\ feed/volume\ of\ feed)(volume\ of\ reactor)}{moles\ A\ feed/time}$

$\qquad\qquad\qquad\qquad = \dfrac{V}{v_o} = \dfrac{reactor\ volume}{volumetric\ feed\ rate\ of\ A}$

Steady-state Mixed Flow Reactor or Continuous-Flow Stirred Tank Reactor (CSTR):

Steady state material balance gives the result

$$\tau = \dfrac{1}{S} = \dfrac{V}{v_o} = \dfrac{VC_{AO}}{F_{AO}} = \dfrac{C_{AO}X_{AO}}{-r_A} \qquad X_A = 0 \quad \text{in feed.}$$

If feed is partially converted i.e $X_A \neq 0$, $\qquad \dfrac{V}{F_{AO}} = \dfrac{X_{Af} - X_{Ai}}{(-r_A)_f}$

Where f and i denote the exit and inlet conditions.

For the special case where the density is constant, $\quad \tau = \dfrac{V}{v} = \dfrac{C_{AO} - C_A}{-r_A}$

Steady-State Plug-Flow Reactor

A steady state material balance yields the relation $V = F_{AO} \int_0^{X_A} \dfrac{dX_A}{-r_A}$

which gives plug-flow reactor volume for conversion X and then $\tau = \dfrac{V}{v_o} = C_{AO} \int_0^{X_A} \dfrac{dX_A}{-r_A}$

If feed is partially converted, $\dfrac{V}{F_{AO}} = \dfrac{V}{C_{AO}v_o} = \int_{X_{Ai}}^{X_{Af}} \dfrac{dX_A}{-r_A} \qquad$ or $\quad \tau = \dfrac{V}{v_o} = C_{AO} \int_{X_{Ai}}^{X_{Af}} \dfrac{dX_A}{-r_A}$

For the special case when density is constant, $\quad \tau = \dfrac{V}{v_o} = - \displaystyle\int_{C_{Ao}}^{C_{Af}} \dfrac{dC_A}{-r_A}$

Mixed-Flow Reactors in series

For a first order reaction system, and equal volume reactors in series, with assumption of constant density, a material balance on reactor i for component A gives

$$\tau = \dfrac{C_o V}{F_o} = \dfrac{V}{v} = \dfrac{C_o \left(X_i - X_{i-1} \right)}{-r_A} \qquad \text{or} \qquad \tau = \dfrac{C_{Ai-1} - C_{Ai}}{K C_{Ai}}$$

Since space-time is the same for all reactors , $\dfrac{C_{Ai-1}}{C_{An}} = (1 + k\tau)^n \quad$ or $\quad C_{An} = C_{AO}(1 + k\tau)^{-n}$

For the reactor system as a whole, $\quad \tau_n = \dfrac{n}{k} \left[\left(\dfrac{C_{AO}}{C_{An}} \right)^{1/n} - 1 \right]$

Example Problems

Kinetics 5-1:

What is the heat of reaction for the following : Given are heats of combustion.

$$C_2H_5OH(l) + O_2(g) \rightarrow CH_3COOH(l) + H_2O(l)$$

C_2H_5OH	$\Delta H_C = - 1366913$ kJ/kg-mol at 25 ^0C
CH_3COOH	$\Delta H_C = - 871695$ kJ/kg-mol at 25 ^0C

Kinetics 5-2:

Ethyl benzene decomposes according to the reaction $C_6H_5C_2H_5 \rightarrow C_6H_5C_2H_3 + H_2$.
The reaction rate constants at two temperatures are as follows

Temp ^0C	$k \times 10^4$
540	1.6
550	2.8

Calculate the activation energy of the reaction.

Kinetics 5-3 :

Calculate free energy change for the reaction :

$$C_2H_5OH(l) + O_2(g) \rightarrow CH_3COOH(l) + H_2O(l)$$

(ΔG_f^0)$_{25}$ (kJ/kgmol) - 174724   -369322   -237191

Kinetics 5-4:

For the reaction $A \rightarrow B$, $\Delta H_R = - 100416$ kJ/kg-mol and $\Delta G_f^0 = -12552$ kJ/kg-mol at 25 ^0C. Calculate the equilibrium constant and equilibrium conversion for the reaction at 60 ^0C. Assume ΔH_R is independent of temperature.

Kinetics 5-5:

A certain reaction has the rate given by $- r_A = 0.01 \, C_A^2$ g-mol/cm^3.min. If the rate is to be expressed in kmol/liter.h , what is the value and units of the rate constant?

Kinetica 5-6:

For a gas phase elementary reaction, A → 3 R, what is the fractional volume change assuming the volume varies linearly between zero and complete conversion. The reaction mixture initially contains 40 % by volume inerts.

Kinetics 5-7:
A substance A decomposes by first order kinetics. In a batch reactor, 50 % A is converted in 5 min.. How much longer would it take to reach 90 % conversion?

Kinetics 5-8:
The first order reversible reaction $A \overset{\leftarrow}{\rightarrow} R$ takes place in a batch reactor. Initial conditions are $C_{RO} = 0$, $C_{AO} = 0.75$ g-mol/liter. After 10 minutes, the conversion of A is 40% and the equilibrium conversion is 65 %. Develop the rate equation for this reaction.

Kinetics 5-9 :
The reaction, $CO_2 + H_2 = CO + H_2O$ is carried out by heating to 1000 K and allowing the reaction to come to equilibrium at 50 bar total pressure. 60 % of CO_2 is found to be converted. What is the value of the equilibrium constant K_P, and what is the partial pressure of CO in the mixture if the initial mixture consisted of only CO_2 and H_2 in equimolar proportion. Assume ideal behavior of components.

Kinetics 5-10:
A liquid phase reaction A → R is carried out in a CFSTR. The reaction rate is 1 mol/liter/h. Feed is 5 % converted (1) Find the size of the reactor needed for 80% conversion if feed is to be processed at a rate of 1200 mols/h. (2) If the feed rate is doubled, what is the size of the reactor needed.

Kinetics 5-11:
A homogeneous liquid phase second-order reaction $2A \rightarrow R$ is carried out in a plug-flow reactor with 60 % conversion. What will be the conversion in a plug-flow reactor 2 times as large if all other variables remain the same, and if the reaction takes place without a volume change? Feed to the reactor is pure A.

Solutions:

Kinetics 5-1:
$$\Delta H_R = (\Sigma \Delta H_c)_{reac\,tan\,ts} - \left(\Sigma \Delta H_C\right)_{products}$$
$$= -1366913 - (-871695) = -495218 \text{ kJ/kg-mol.}$$

Kinetics 5-2:
$$Arrheneous\ relation \quad k = ae^{-E/RT}$$
$$T_2 = 550 + 273 = 823 \text{ K} \qquad T_1 = 540 + 273 = 813 \text{ K}$$

$$\frac{k_2}{k_1} = \frac{e^{-E/RT_2}}{e^{-E/RT_1}}$$

By taking logarithms $\quad \ln\frac{k_2}{k_1} = -\frac{E}{RT_2} + \frac{E}{RT_1}$

From which, $E = \ln(k_2/k_1)\left[R\left(\frac{T_2T_1}{T_2 - T_1}\right)\right]$
$$= \ln\left(\frac{2.8\times10^{-4}}{1.6\times10^{-4}}\right)\left[8.314 \times \frac{823\times813}{823-813}\right] = 311308 \text{ kJ/kg-mol.}$$

Kinetics 5-3:

$$\Delta G_f^0 = -369322 - 237191 + 174724 = -431789 \text{ kJ/kg-mol}$$

Kinetics 5-4:

ΔH_R is independent of temperature. Therefore $\ln \dfrac{K_2}{K_1} = -\dfrac{\Delta H_R}{R}\left[\dfrac{1}{T_2} - \dfrac{1}{T_1}\right]$

Also, $\quad \Delta G_f^0 = -RT \ln K \quad \ln K = -\Delta G_f^0/RT$

$$K = e^{-\Delta G_f^0/RT} = e^{-(-12552)/(8.314 \times 298)} = 158.6 \text{ m} \quad T_2 = 60 + 273 = 333 \text{ K}$$

$$\ln \frac{K_2}{158.6} = \frac{-100416}{8.314}\left[\frac{1}{333} - \frac{1}{298}\right] = -4.26 \text{ Therefore,}$$

$$K_2/158.63 = e^{-4.26} = 0.014122$$

and then $K_2 = 0.014122 \times 158.62 = 2.24$

At equilibrium, $\quad K = \dfrac{X_{AE}}{1-X_{AE}} = 2.24$

Therefore, $\quad X_{AE} = K/(K+1) = 0.6915$

Kinetics 5-5:

$$-r_A = 0.01\, C_A^2 \qquad \frac{g\text{-}mol}{cm^3\,min}$$

$$= 0.01 C_A^2 \quad \frac{\dfrac{g\text{-}mol}{1000}\dfrac{kmol}{g\text{-}mol}}{\dfrac{cm^3}{1000}\dfrac{liter}{cm^3}\,\dfrac{min}{60}\dfrac{h}{min}} = 0.01 \times 60 \quad \frac{kmol}{liter.h} = 0.6\, C_A^2$$

kmol/(liter.h)

The value of rate constant is 0.6

Since $\quad k = -r_A/C_A^2$, the units of k are now $\quad k = \dfrac{kmol\,(/liter.h)}{(kmol\,/\,liter)^2} = \dfrac{liter}{kmol.h}$

Kinetics 5-6:

$$A \to 3R$$

	initial	final
A	0.6	0.0
R	0.0	1.8
Inerts	0.4	0.4
Total	1.0	2.2

Therefore, $\quad \epsilon_A = \dfrac{2.2 - 1.0}{1.0} = 1.2$

Kinetics 5-7:

In terms of conversion, $\quad -\ln(1-X_A) = kt \quad$ solution of first order rate equation.

At t = 5 min. $X_A = 0.5 \quad$ Therefore, $-\ln(1 - 0.5) = k \times 5$

Then k $= -\ln 0.5/5 = 0.13863 \text{ min}^{-1}$

For 90 % conversion, $-\ln(1 - 0.9) = 0.13863 \times t$

$t = -\ln(0.1)/0.13683 = 16.6 \text{ min}$

Therefore, extra time required to reach 90 % conversion = 16.6 - 5.0 = 11.6 min.

Kinetics 5-8:

Rate equation $\quad \dfrac{dX_A}{dt} = \dfrac{k_1(M+1)}{M+X_{Ae}}\left(X_{Ae} - X_A\right)$

In terms of conversion, $\quad -\ln\left(1 - \dfrac{X_A}{X_{Ae}}\right) = \dfrac{M+1}{M+X_{Ae}}k_1 t$

In this case, $M = C_{Ro}/C_{Ao} = 0$

Therefore, $-\ln(1 - X_A/X_{Ae}) = (1/X_{Ae})k_1 t$

$X_{Ae} = 0.65$, and $X_A = 0.4$ when t = 10 min.

$-\ln\left(1 - \frac{0.4}{0.65}\right) = (1/0.65) \times k_1 \times (10)$ from which $k_1 = 0.0621$ min⁻¹

Rate equation is $\frac{dX_A}{dt} = \frac{k_1(M+1)}{M+X_{Ae}}\left(X_{Ae} - X_A\right) = \frac{0.0621(0+1)}{(0+0.65)}\left[0.65 - X_A\right]$

Therefore, rate equation is $\frac{dX_A}{dt} = 0.0621 - 0.09554 X_A$

Kinetics 5-9:

$$K_P = \frac{y_{CO} y_{H_2O}}{y_{CO_2} y_{H_2}} \times \frac{50 \times 50}{50 \times 50} = K_y$$

	initial	at equilibrium
CO₂	1.0	0.4
H₂	1.0	0.4
CO	0.0	0.6
H₂O	0.0	0.6
	2.0	2.0

$y_{CO} = \frac{0.6}{2} = 0.3$ $y_{H_2O} = 0.3$ $y_{CO_2} = 0.4/2 = 0.2$ $y_{H_2} = 0.2$

$K_P = \frac{0.3 \times 0.3}{0.2 \times 0.2} = 2.25$ and partial pressure of CO = 0.3(50) = 15 barA.

Kinetics 5-10:

(1) For a CFSTR, $\frac{V}{F_{Ao}} = \frac{X_{Af} - X_{Ai}}{(-r_A)_f}$

$V = \frac{X_{Af} - X_{Ai}}{(-r_A)_f} \times F_{Ao} = \frac{0.8 - 0.05}{1} \times 1200 = 900$ liters

(2) V = 0.75(1200x2) = 1800 liters.

Kinetics 5-11:

For no volume change, $\varepsilon = 0$, $-r_A = -\frac{dC_A}{dt} = C_{AO}\frac{dX_A}{dt} = kC_{AO}^2(1-X_A)^2$

For a plug flow reactor, $V = F_{A0}\int_0^{X_A}\frac{dX_A}{-r_A} = F_{A0}\int_0^{X_A}\frac{dX_A}{kC_{AO}^2(1-X_A)^2} = \frac{F_{A0}}{kC_{AO}^2}\left[\frac{1}{1-X_A}\right]_0^{X_A}$

$= \frac{F_{A0}}{kC_{AO}^2}\frac{X_A}{1-X_A}$

All conditions except the volume are given to be the same, therefore by comparison of the two reactor volumes, the following can be written

$$\frac{V_2}{V_1} = \frac{[X_A'(1-X_A)]_2}{[X_A'(1-X_A)]_1}$$

or

$$2 = \frac{\left[X_A \,/\, \left(1 - X_A\right) \right]_2}{[0.6 \,/\, (1 - 0.6)]_1}$$

Then

$$\left[\frac{X_A}{1 - X_A} \right]_2 = 2 \times \left[\frac{0.6}{1 - 0.6} \right]_1 = 3$$

Solving,

$$[X_A]_2 = 0.75$$

Therefore by doubling the reactor volume but keeping the same other conditions, 75 % conversion will be obtained.

References:

1. Das D.K. and R.K. Prabhudesai, **Chemical Engineering License Review**, Engineering Press, 2nd Ed.,(1996)
2. Smith J. M., **Chemical Engineering Kinetics**, McGraw-Hill, New York, 3rd Ed. (1981)
3. AIChE Modular Instruction Series, **Kinetics**, vol 1-5
4. Levenspiel Octave, **Chemical Reaction Engineering**, John Wiley & Sons, 2nd Ed., (1972)

Chapter 6
PROCESS DESIGN AND ECONOMIC EVALUATION

The best equipment design is the least expensive and safe design that meets the specification and performs the duty. Since all businesses are motivated by profit, economic evaluation plays an important role in process design. There are many ways to reach the same destination. The best way is the least expensive way in terms time, money, and safety.

Degrees of Freedom for Optimization, F.

It is the number of variables which are not fixed or specified in a design, and therefore, are free to be adjusted for optimum design.

$$F = M - N$$

where:
M = number of variables, such as equipment type, pressure, temperature etc.,
N = number of <u>independent</u> sources of information, such as heat and material balances.

The degrees of freedom, F, has two components.
$$F = F_1 + F_2$$
where:
F_1 = environmental degrees of freedom : variables whose values are fixed by the environment of the process, such as cooling water supply temperature or the inlet and outlet temperatures of process side of a heat exchanger etc. This is a freedom lost by the optimizer.
F_2 = economic degrees of freedom : variables that the designer can adjust to maximize the profitability, such as the type of an exchanger or the cooling water flow rate etc.

Problem 6.1 The following table lists the variables of a heat exchanger handling fluids with no change of phase:

	Hot Side		Cold Side
Flow rate	W_1		W_2
Temperature in	T_1		T_2
Temperature out	T_3		T_4
Heat exchanged		Q	
Area		A	
Logarithmic Temperature Difference		LMTD	
Overall Heat Transfer Coefficient		U	
Type of Exchanger		K	

Find the degrees of freedom for optimization. If W_1, T_1, T_3, T_2, and K are fixed by the environment, and find the number of economic degrees of freedom.

Fixed Investment, I_F
This is the investment associated with all processing equipment and auxiliary facilities that are needed for the complete operation.

Gross Profit, R

$$R = S - C$$

Where:
R = gross profit $/year
S = sales revenue, $/year
C = cost of production, $/year

Net Profit, P_N

$$P_N = (R - dI_F)(1-t)$$

where:
d = depreciation, decimal
t = tax rate, decimal

Return On Investment (ROI)

Return on investment is generally defined by:

$$ROI = \frac{Average\ annual\ after\text{-}tax\ profit\ during\ earning\ life}{Original\ fixed\ investment\ +\ working\ capital}$$

If working capital is ignored, ROI may be expressed by:

$$ROI = \frac{(R - dI_F)(1 - t)}{I_F}$$

Very often, an incremental investment is necessary to reduce annual operating cost. When the sales revenue is the same, the return on such incremental investment is justified by:

$$ROI = \frac{[(C + dI_F)_{basecase} - (C + dI_F)_{othercase}](1 - t)}{(I_F)_{othercase} - (I_F)_{basecase}}$$

Payout or Payback Time, T

$$T = \frac{I_F}{R-(R-dI_F)t}$$

Often, for simplicity, taxes are not considered, and $T = I_F/R$

Venture Profit, V

$$V = (R - dI_F)(1 - t) - i_m(I_F + I_w)$$

Venture Cost, V_c

$$V_c = (C + dI_F)(1 - t) + i_m(I_F + I_w)$$

where:
i_m = minimum acceptable rate of return, decimal
I_w = working capital: money invested in raw materials etc., and cash required to run the project. This may be ignored for simplified analysis.

Optimization study generally involves maximization of venture profit , minimization of venture cost or total annual cost. Total annual cost includes fixed annual cost and variable annual cost. The fixed annual cost is independent of production, and includes such items as depreciation on fixed investment, minimum allowable return on investment after income taxes, taxes and insurance, cost of safety, administration, general maintenance, and general services. The variable annual cost includes cost of raw materials, direct labor, utilities, maintenance cost dependent on operation, warehouse and shipping.

Problem 6.2 You have been asked to reduce the cost of solvent losses by installing a refrigerated condenser. Your company investment guideline requires an ROI after tax greater than 25% for this type of investment. From the following data, decide whether the installation is justified or not. Depreciation of fixed investment is 10%, and tax rate is 52%.

	Base Case	**Proposed Project**
Investment	0	$200,000
Annual cost		
(A) Solvent losses	$400,000/yr	$100,000/yr
(B) Power cost	$200,000/yr	$225,000/yr
(C) Cooling water	$90,000/yr	$100.000/yr
Total	$690,000/yr	$425,000/yr

Linear Break-Even Analysis

$$Q_B = \frac{F}{s - c}$$

where:

Q_B = break-even capacity, unit/time
F = Fixed cost, \$/time
s = selling price, \$/unit
c = variable cost, \$/unit

Method of making order-of-magnitude estimate of equipment cost and installed cost
The cost of an equipment may be expressed in the form:

$$C=K(Size)^n$$

where:

C = equipment cost
K, n are constants related to the equipment type.
The approximate installed cost of the equipment may be obtained by multiplying the cost of the equipment by a factor(Lang factor) as shown in the following table. The value of K and n can be estimated from costs of two sizes. Given the cost of one size, the cost of another size may be estimated from the following table.

Equipment	Size Basis	Value of exponent, n	Lang Factor,K
Pump	HP	0.52	4.0
Shell & Tube			
Heat exchanger	Area	0.62	3.3(ss), 4.0(cs)
Air cooler	Area	0.75	2.2(ss), 2.3(cs)
Column	Diameter	0.65	4.0
Crystallizer	Tons/day dry solid	0.55	2.4
Plate & frame filter	Area	0.58	2.7
Rotary vacuum filter	Effective area	0.63	2.4
Rotary kiln dryer	Area	0.8	3.0
Blower & fan	Cfm	0.68	2.8
Bowl centrifuge	HP	0.73	2.4
Boiler	Lbs/h	0.7	1.8
Ball Mills	Tons/h	0.7	1.8
Pulverizer	Tons/h	0.39	2.5
Atm. Storage tank	Gallon	0.6	4.0
Pressure vessel	Gallon	0.6	2(ss,gl), 4(cs)
Agitator	HP	0.56	As above
Cooling tower	Gpm	0.6	1.8
Refrigeration(mech)	Tons	0.55	1.4
Compressor	HP	0.8	3.1

Legend: ss = stainless steel, cs = carbon steel, gl = glass lined.

Problem 6.3 The cost of a 250 m² exchanger is $500,000. What is the estimated order-of-magnitude cost of a similar 900 m² exchanger? What will be the installed cost of the 900 m² exchanger? Use 0.62 exponent and 3.3 as the Lang factor.

Problem 6.4 Formulate the equation of the total annual cost of insulating an oven of 10 m² heat transfer area as a function of insulation thickness, T. The following information is available:

Inside wall temperature:	293 °C
Outside temperature:	15 °C
Air film heat transfer coefficient:	23 w/.m². K
Thermal conductivity of insulation:	0.05 w/m².k/mv
Installed insulation cost:	$425/m³
Cost of heat:	$.004/10⁶J
Economic life:	10 years
Annual operating hours:	8700 hrs
Maintenance:	3% of fixed investment
Insurance and property taxes:	1.5% of fixed investment

SOLUTIONS

Problem 6.1

The number of variables, M, given by the problem: 11
Compute the number of independent sources of information:

$$Q = UA(\Delta T)_{LMTD}$$

$$U = Function(W,T,K)$$

$$\Delta T_{LMTD} = \frac{(T_1 - T_4) - (T_3 - T_2)}{\ln\dfrac{(T_1 - T_4)}{(T_3 - T_2)}}$$

$$Q = W_1 C_{p1}(T_3 - T_1)$$

$$W_1 C_{p1}(T_1 - T_3) = W_2 C_{p2}(T_4 - T_2)$$

Total number of independent sources of information = 5
Total number of freedom for optimization = 11 - 5 = 6.
Total number of environmental degrees of freedom, by the problem, is 5.
Hence, total number of economic degrees of freedom = 6 - 5 = 1.

Problem 6.2

The justification of the project should be based on ROI of the incremental cost of the proposed project relative to the base case.

$$ROI = \frac{[(C + dI_F)_{base\ case} - (C + dI_F)_{other\ case}](1 - t)}{(I_F)_{other\ case} - (I_F)_{base\ case}}$$

$$Hence,\ ROI = \frac{[690000 - (425000 + 0.1 x 200000)](1 - 0.52)}{200000}$$

$$= 58.8\% > 25\%$$

Therefore, the installation is economically justified.

Problem 6.3

The estimated cost of 900 m^2 exchanger = 500000x(900/250)$^{0.62}$ = $1,106,312

Installed cost = 3.3x 1106312 = $3,650,830

Problem 6.4

Investment for insulation = (10 m^2)x(T meter)x(425 $/m^3) = 4,250 T($)

$$Heat\ lost = UA(\Delta t) = \frac{A(\Delta t)}{\frac{1}{h}+\frac{T}{k}} = \frac{(10)(293-15)}{0.04348+20T} = \frac{2780}{0.04348+20T}\ \ W$$

$$Annual\ cost\ of\ heat\ lost\ =\ \frac{2780x8700x3600x10^{-9}}{0.04348+20T} = \frac{348.28}{0.04348+20T}$$

Annual depreciation = 0.1x4250T = 425T

Annual maintenance cost = 0.03x4250T = 127.5T

Annual taxes and insurance = 0.015x4250T = 63.75T

$$Total\ annual\ cost\ =\ \frac{348.28}{0.04348+20T}+616.25T$$

Chapter-7
Heat Transfer
SI units are noted in [ ]

Three modes of heat transfer are conduction, convection, and radiation.

CONDUCTION
It is a mode of heat transfer across a medium caused by temperature difference.
A steady state conduction of heat is given by Fourier's law:

$$q_k = -kA(dT/dx) \tag{7-1}$$

where q_k = heat transferred by conduction in the x direction, Btu/h[W],
$\quad$ k = thermal conductivity of material, Btu.ft/h.ft^2.°F[W/m.K],
$\quad$ A = area perpendicular to the direction of heat flow, ft^2[m^2]
(dT/dx) = temperature gradient in the x direction, °F/ft[K/m]

Application of Fourier's law to simple geometries

(a) Flat single plate

$$q_k = kA(T_{hot} - T_{cold})/X = (T_{hot} - T_{cold}) = \Delta T/R_k \tag{7-2}$$

where X is the plate thickness, ft[m], and R_k = X/kA is the thermal resistance, h.°F/Btu[K/W].

(b) Composite Flat Walls

The rate of heat transfer through a series of flat plates forming a composite wall is given by:

$$q_k = \frac{T_0 - T_1}{R_{k1}} = \frac{T_1 - T_2}{R_{k2}} = .. = \frac{T_{n-1} - T_n}{R_{kn}} = \frac{T_0 - T_n}{R_{kt}} \tag{7-3}$$

where $T_0 > T_1 > ...> T_n$, R_{kn} = resistance of nth plate = $X_n/(k_n A_n)$,
and R_{kt} = total resistance = $(R_{k1} + R_{k2} + ... + R_{kn})$

(c) Cylindrical Pipe

$$q_k = \frac{2\pi kL(T_i - T_o)}{\ln(r_o/r_i)} = \frac{T_i - T_o}{\ln(r_o/r_i)/(2\pi kL)} = \frac{T_i - T_o}{R_k} \tag{7-4}$$

where L = length of pipe, ft[m]; T_i = temperature at the inside wall, °F[K];
T_o = temperature at the outside wall, oF[K]; r_o, r_i = outside, inside radii of pipe, both in the same unit,
$R_k = \ln(r_o/r_i)/(2\pi kL)$,h.°F/Btu[K/W]. The heat transfer through composite cylindrical walls may be

computed in the same manner as shown in equation (7-3) but using the resistance formula noted in this paragraph.

(d) Hollow Sphere.

$$q_k = \frac{T_i - T_o}{(r_o - r_i)/(4\pi k r_o r_i)} = \frac{T_i - T_o}{R_k} \qquad (7\text{-}5)$$

where r_o, r_i are the radii of the outside and inside surfaces, both in the same unit. The heat transfer through a composite sphere may be computed in the same manner as shown in equation (7-3) but using the resistance formula noted in equation (7-5).

CONVECTION

It is a mode of heat transfer between a surface and a moving fluid when they are at different temperatures. When the fluid moves naturally, induced by buoyancy, the mode of heat transfer is called natural convection, otherwise it is called as forced convection. It is generally assumed that the resistance to convective heat transfer lies in the boundary layer via a conductive mode. By analogy with equation (7-2):

$$q_c = kA(T_{hot} - T_{cold})/X_{boundary\ layer} = h_c A(T_{hot} - T_{cold})$$

$$= \frac{T_{hot} - T_{cold}}{1/(h_c A)} \qquad (7\text{-}6)$$

where q_c is the convective heat transfer rate, Btu/h[W], h_c is the convective heat transfer coefficient, Btu/h.ft^2.°F[W/m^2.K].

OVERALL HEAT TRANSFER COEFFICIENT, U, THROUGH COMPOSITE WALLS INCLUDING INSIDE AND OUTSIDE CONVECTION

$$1/U = 1/h_i + \Sigma X_i/k_i A_i + 1/h_o \qquad (7\text{-}6a)$$
$$q = UA\Delta T \qquad (7\text{-}6b)$$

RADIATION

It is a mode of heat transfer that requires no medium in contrast with the two modes described earlier. Radiant energy is emitted in the form of electromagnetic waves or photons from any matter(solid, liquid, or gas), which is at a finite temperature at the expense of its internal energy.

Of the total radiant energy received by a body, a fraction(α = absorptivity) is absorbed by the body, a fraction(R = reflectivity) is reflected away from the body, and the balance(τ= transmittivity) is transmitted through the body. Thus:

$$\alpha + R + \tau = 1 \tag{7-7}$$

For a black body, $R = \tau = 0$, so $\alpha = 1$
For an opaque body $\tau = 0$, so $\alpha + R = 1$

Radiation from a body

The radiant energy emitted from a surface is given by Stefan-Boltzmann law:

$$q_{r1-} = \sigma A_1 \varepsilon_1 T_1^4 \tag{7-8}$$

where q_{r1-} = heat emitted due to radiation from body 1, Btu/h[W]; σ = Stefan-Boltzmann constant = 0.173×10^{-8} Btu/(h.ft^2.°R^4)[5.67×10^{-8} W/m^{2}K^4]; A_1 = surface area of the body, ft^2[m^2]; ε_1 = emissivity of body 1(depends on the nature of the surface), fraction; T_1 = temperature of body 1, °R[K].

Radiation onto a body

The radiant energy absorbed by a body(1) of surface of area A_1 at temperature T_1 from black body(2) surroundings or enclosure at T_2 may be given by:

$$q_{r1-2} = \sigma A_1 \alpha_{1-2} T_2^4 \tag{7-9}$$

where the arrow denotes the direction of the flow of radiant energy. α_{1-2} is the ratio of the energy absorbed by the body(1) to the energy incident coming from the enclosure at T_2.

Radiation exchange between a body of area A_1 and temperature T_1 and its enveloping surrounding at temperature T_2

$$q_{r12} = \sigma A_1 (\varepsilon_1 T_1^4 - \alpha_{1-2} T_2^4) \tag{7-10}$$

Black body
A black body is an ideal surface having $\alpha = \varepsilon = 1$. In real situation, when a small object is enclosed in a large chamber, the chamber may be treated as a black body.

Graybody
For graybody, $\alpha = \varepsilon$ = constant and <1 at all temperature. Many real objects behave like graybodies. Radiation between two close, adjacent gray surfaces, each of area A, but at temperatures T_1 and T_2 may be given by:

$$q_{r12} = \frac{\sigma A(T_1^4 - T_2^4)}{1/\varepsilon_1 + 1/\varepsilon_2 - 1} \tag{7-11}$$

Radiation between one gray surface(A_1, T_1, ε_1) enclosed by another gray surface (A_2, T_2, ε_2) may be given by:

$$q_{r12} = \frac{\sigma A_1(T_1^{\,4} - T_2^{\,4})}{1/\varepsilon_1 + (A_1/A_2)(1/\varepsilon_2 - 1)} \tag{7-12}$$

When $A_2 \gg A_1$:

$$q_{r12} = \sigma A_1 \varepsilon_1 (T_1^{\,4} - T_2^{\,4}) \tag{7-13}$$

When A_1 is in ft^2, and T is in °R:

$$q_{r12} = 0.173 A_1 \varepsilon_1 [(T_1/100)^4 - (T_2/100)^4], \text{ Btu/h} \tag{7-14}$$

$$h_r = \frac{0.173\varepsilon_1[(T_1/100)^4 - (T_2/100)^4]}{T_1 - T_2} \tag{7-15}$$

where h_r is radiation heat transfer coefficient, Btu/h.ft^2.°F.

When A_1 is in m^2, and T is in K:

$$q_{r12} = 5.67 A_1 \varepsilon_1 [(T_1/100)^4 - (T_2/100)^4], \text{ W} \tag{7-16}$$

$$h_r = \frac{5.67\varepsilon_1[(T_1/100)^4 - (T_2/100)^4]}{T_1 - T_2} \tag{7-17}$$

where h_r is radiation heat transfer coefficient, W/m^2.K

**HEAT TRANSFER THROUGH FLUID TO FLUID ACROSS A CYLINDRICAL WALL
WITH FOULING ON BOTH SIDES. MODES OF HEAT TRANSFER FROM INSIDE OUT
VIA CONVECTION, CONDUCTION, AND COMBINED CONVECTION & RADIATION.**

$$q = \frac{T_1-T_2}{1/(h_{12}A_3)} = \frac{T_2-T_3}{f_{d23}/A_3} = \frac{T_3-T_4}{l_{34}/(k_wA_{mp})} = \frac{T_4-T_5}{f_{d45}/A_4}$$

$$= \frac{T_5-T_6}{1/(h_{56}+h_r)A} = \frac{T_1-T_6}{1/(U_3A_3)} = \frac{T_1-T_6}{1/(U_4A_4)}$$

(7-18)

Let $T_1-T_6 = \Delta T$, $h_{12} = h_i$, $A_3 = A_i = \pi D_iL$, $f_{d23}=f_{di}$, $l_{34}=l_w$, $A_{mp}=\pi((D_o+D_i)/2)L=\pi D_{av}L$,
$f_{d45}=f_{do}$, $A_4=A_o=\pi D_oL$, $h_{56}=h_o$, $U_3=U_i$, $U_4=U_o$:

$$q = U_iA_i(\Delta T) = U_oA_o(\Delta T)$$

(7-19)

$$\frac{1}{U_iA_i} = \frac{1}{U_oA_o}$$

(7-20)

$$\frac{1}{U_o} = \frac{D_o}{h_iD_i} + \frac{f_{di}D_o}{D_i} + \frac{l_wD_o}{k_wD_{av}} + f_{do} + \frac{1}{h_o+h_r}$$

(7-21)

Where U = overall heat transfer coefficient, Btu/h.ft².°F[W/m².K], h = local convection heat transfer
coefficient, h_r = radiation heat transfer coefficient, D = diameter of cylinder, f_d = fouling resistance,
h.ft².°F/Btu[m².K/W], l_w = thickness of cylinder wall, ft[m], k_w = thermal conductivity of
wall,Btu/h.ft.°F [W/m.K], suffix(i=inside, o=outside).

When surfaces are clean, inside coefficient is high, wall resistance is low:

$$U_o = h_o + h_r$$

(7-22)

When surfaces are clean, wall resistance is low, radiation coefficient is negligible, and D_o/D_i is
assumed 1 for approximation:

$$U_o = (h_o)(h_i)/(h_o + h_i)$$

(7-23)

LOG MEAN TEMPERATURE DIFFERENCE(LMTD) AND CORRECTION FACTOR(F_T)

Countercurrent   **Cocurrent**

T_1---hot fluid--->T_2 T_1--hot fluid--->T_2
t_2<--cold fluid---t_1 t_1--cold fluid-->t_2

$\Delta T_1 = T_1 - t_2$ $\Delta T_2 = T_2 - t1$ $\Delta T_1 = T_1 - t_1$ $\Delta T_2 = T_2 - t_2$

$$LMTD = \frac{\Delta T_1 - \Delta T_2}{\ln[(\Delta T_1)/(\Delta T_2)]}$$ (7-24)

Obviously, when terminal temperature differences are identical, LMTD does not apply, and the terminal temperature difference is used as the driving force.

Unless one or both of the fluids exchange heat under isothermal condition, LMTD for countercurrent flow for the same process terminal temperatures will be greater than the LMTD for cocurrent flow, and therefore, LMTD for countercurrent flow has greater potential for heat recovery. When at least one fluid exchanges heat isothermally, LMTD for countercurrent flow is the same as that in the cocurrent flow.

In the multipass shell and tube heat exchangers, the flows are not truly countercurrent. Therefore, a correction factor, F_T is introduced. See reference[4(a)] for evaluation of this parameter. The value of F_T is 1 for true countercurrent exchanger, and is less than 1 for any other design, implying the higher heat recovery by a countercurrent arrangement than any other arrangement for non-isothermal exchange of heat.

$$q = UAF_T(\Delta T)_{lm}$$ (7-25)

ESTIMATION OF OUTLET TEMPERATURES FOR COCURRENT AND COUNTERCURRENT FLOWS WITH NO PHASE CHANGE, GIVEN INLET TEMPERATURES(T_1,t_1), SPECIFIC HEATS(C, c),OVERALL HEAT TRANSFER COEFFICIENT,U, HEAT TRANSFER AREA,A, AND FLOW RATES(W,w).

For hot fluid:	$q = WC(T_1-T_2)$	(7-26)
For cold fluid:	$q = wc(t_2-t_1)$	(7-27)
Heat balance:	$q = WC(T_1-T_2) = wc(t_2-t_1) = UA(\Delta T)_{lm}$	(7-28)

Define $R = wc/(WC)$ and $K_1 = e^{UA(R-1)/wc}$, $K_2 = e^{UA(R+1)/wc}$, then:
For countercurrent flow:

$$T_2 = \frac{(1-R)T_1 + (1-K_1)Rt_1}{1-RK_1} \qquad (7-29)$$

For cocurrent flow:

$$T_2 = \frac{(R+K_2)T_1 + (K_2-1)Rt_1}{(R + 1)K_2} \qquad (7-30)$$

For both cases, calculate t_2 from equation (7-28):

$$t_2 = (T_1-T_2)/R + t_1 \qquad (7-31)$$
$$= (T_1-t_1)(K_1-1)/(RK_1-1) + t_1 \qquad (7-31a)$$

Symbols of above equations are as follows:

	Hot Fluid	**Cold fluid**
Flowrate	W	w
Temperature in	T_1	t_1
Temperature out	T_2	t_2
Specific heat	C	c

UNSTEADY STATE HEAT TRANSFER

1. To calculate the time to heat a liquid mass in a perfectly mixed tank heated by an **isothermal heating medium** flowing through a coil or jacket(agitator heat or other heat effect is not considered).

$$\Theta = \frac{Mc}{UA} \ln \frac{T_1-t_1}{T_1-t_2} \qquad (7-32)$$

$$t_\Theta = T_1 - (T_1 - t_1)e^{-k\Theta}, \text{ where } k = UA/(Mc) \tag{7-33}$$

$$Q_\Theta = UA(T_1 - t_1)e^{-k\Theta} \tag{7-34}$$

Here, Θ = heating time, h[s]; M = mass of liquid to be heated, lb[kg]; c = specific heat of the liquid to be heated, Btu/lb.°F[J/kg.K]; U = overall heat transfer coefficient of the heating surface, Btu/h.ft^2.°F[W/m^2.K]; A = heat transfer area, ft^2[m^2]; T_1 = constant temperature of heating medium, °F[K]; t = batch temperature, °F[K], subscript(1=initial, Θ = at time Θ hr[s], 2=final); Q = instantaneous heat load, Btu/h[W].

2. To calculate the time to cool a liquid mass in a perfectly mixed tank cooled by an **isothermal cooling medium** flowing through a coil or jacket(agitator heat or other heat effect is ignored).

$$\Theta = \frac{MC}{UA} \ln \frac{T_1 - t_1}{T_2 - t_1} \tag{7-35}$$

$$T_\Theta = t_1 + (T_1 - t_1)e^{-k\Theta}, \text{ where } k = UA/(MC) \tag{7-36}$$

$$-Q_\Theta = UA(T_1 - t_1)e^{-k\Theta} \tag{7-37}$$

Here, Θ = cooling time, h[s]; M = mass of liquid to be cooled, lb[kg]; C = specific heat of the liquid to be cooled, Btu/lb.°F[J/kg.K]; U = overall heat transfer coefficient of the heating surface, Btu/h.ft^2.°F[W/m^2.K]; A = heat transfer area, ft [m̂]; t_1 = constant temperature of cooling medium, °F[K]; T = batch temperature, °F[K], subscript(1=initial, Θ = at time Θ hr[s], 2=final); Q = instantaneous cooling load, Btu/h[W].

3. To calculate the time to heat a liquid mass in a perfectly mixed tank heated by a **non-isothermal heating medium** flowing through a coil or jacket(agitator heat or other heat effect is not considered).

$$\Theta = \frac{Mc}{WC} \times \frac{K}{K-1} \ln\frac{T_1 - t_1}{T_1 - t_2} \tag{7-38}$$

$$K = e^{UA/(WC)}$$

$$t_\Theta = T_1 - (T_1 - t_1)e^{-K'\Theta}, \text{ where } K' = \frac{WC(K-1)}{McK} \tag{7-39}$$

$$Q_\Theta = \frac{WC(K-1)(T_1 - t_1)e^{-K'\Theta}}{K} \tag{7-40}$$

where T_1 is the inlet temperature of heating medium, °F[K]; W is the flow rate of heating medium, lb/h[kg/s]; C is the specific heat of heating medium, Btu/lb.°F[J/kg.K]; t is the batch temperature, subscript(1= initial, Θ = at time Θ hr[s], 2=final)

4. To calculate the time to cool a liquid mass in a perfectly mixed tank cooled by a **non-isothermal cooling medium** flowing through a coil or jacket(agitator heat or other heat effect is ignored).

$$\Theta = \frac{MC}{wc} \times \frac{K}{K-1} \ln \frac{T_1-t_1}{T_2-t_1} \qquad (7-41)$$

$$K = e^{UA/(wc)}$$

$$T_\Theta = t_1 + (T_1 - t_1)e^{-K'\Theta}, \text{ where } K' = \frac{wc(K-1)}{MCK} \qquad (7-42)$$

$$-Q_\Theta = \frac{wc(K-1)(T_1-t_1)e^{-K'\Theta}}{K} \qquad (7-43)$$

where t_1 is the inlet temperature of cooling medium, °F[K]; w is the flow rate of cooling medium, lb/h[kg/s]; c is the specific heat of cooling medium, Btu/lb.°F[J/kg.K]; T is the batch temperature, subscript(1= initial, Θ = at time Θ hr[s], 2=final).

References:

1. "*Chemical Engineering License Review*", D. K. Das & R. K. Prabhudesai, Engineering Press, Austin, Texas

2. "*Engineering Flow and Heat Transfer*", O. Levenspiel, Plenum Press, NY

3. "*Fundamentals of Heat transfer*", F. P. Incropera & D. P. Dewitt, John Wiley & Sons, NY

4. "*Perry's Chemical Engineers' Handbook*", R.H.Perry, D. Green,McGraw-Hill, Inc, N.Y., 6th edition:(a)p 10-27

5. G. Goff, Personal communications.

Problems
Heat Transfer

Problem 7.1 A countercurrent shell-and-tube exchanger has the following process data:

	Tube side	Shell side
Flow, lb/h	13,000	8,000
Temperature in, °F	110	510
Specific heat, Btu/lb.°F	1.2	0.9

Outside heat transfer area, ft²: 400
Overall heat transfer coefficient, Btu/h.ft².°F: 100
Estimate the tube side outlet temperature.

Problem 7.2 A batch is to be heated in a coil-in-tank agitated vessel using saturated steam. The following data are available:

Batch size:	45000 lbs
Overall heat transfer coefficient:	50 Btu/h.ft².°F
Heating area of coil immersed:	900 ft²
Average specific heat of batch:	0.5 Btu/lb.°F
Initial Batch temperature:	100°F
Steam temperature:	448°F
Latent heat of steam:	777 Btu/lb

Calculate the batch temperature and instantaneous steam consumption rate after 30 minutes.

HEAT TRANSFER
SOLUTIONS

Problem 7.1

From equation (7-29):

$R = wc/WC = 13000 \times 1.2/8000 \times 0.9 = 2.167$

$K_1 = e^{UA(R-1)/wc} = e^{100 \times 400(2.167-1)/13000 \times 1.2} = 19.932$

$$T_2 = \frac{(1-2.167)510 + (1-19.932)2.167 \times 110}{1-2.167 \times 19.932} = 121.06$$

From equation (7-31):

$t_2 = (510-121.06)/2.167 + 110 = 289.5$ °F **Answer**

Alternatively, equation (7-31a) may be used, thereby avoiding calculation of T_2: and saving time:

$t_2 = (T_1-t_1)(K_1-1)/(RK_1-1) + t_1$
$= (510-110)(19.932-1)/(2.167 \times 19.932-1) + 110 = 289.5$

Problem 7.2

From equation(7-33):

$t_\Theta = T_1 - (T_1-t_1)e^{-k\Theta}$

$k = UA/Mc = 50 \times 900/45000 \times 0.5 = 2$
$t_\Theta = 448 - (448-100)e^{-2 \times 0.5} = 319.98$ °F

From equation (7-34):

$Q_{\Theta=0.5} = UA(T_1-t_1)e^{-k\Theta} = 50 \times 900(448-100)e^{-1} = 5760992$ Btu/h

Steam consumption $= 5760992/777 = 7414.4$ lb/h

Chapter - 8
TRANSPORT PHENOMENA
SI units are noted in [ ].

Transport phenomena deal with the transfer of momentum, mass, and heat in a fluid. In this chapter, we will deal with the fundamentals of momentum transfer only.

DENSITY, SPECIFIC VOLUME, SPECIFIC WEIGHT, AND SPECIFIC GRAVITY

Density(ρ) = mass/volume, lb/ft^3 [kg/m^3]
Specific volume(υ) = 1/density, ft^3/lb [m^3/kg]
Specific weight(γ) = weight/volume = mass x acceleration due to gravity/volume
$\qquad\qquad$ = $\rho g/g_c$, lbf/ft^3[N/m^3]
Specific gravity of substance(s) = density of substance/density of water at 4°C
$\qquad\qquad\qquad$ = ρ(lb/ft^3)/62.43
$\qquad\qquad\qquad$ = ρ(kg/m^3)/1000
Specific gravity of gas relative to air = Molecular weight of gas/28.97

Notes:
g = acceleration due to gravity, 32.17 ft/s^2[9.81 m/s^2], varies from place to place, and planet to planet.
g_c = Newton's-law proportionality factor for the gravitational force unit
$\quad$ = 32.17 lb.ft/s^2.lbf[1 kg.m/s^2.N]. This is a constant factor.
Specific gravity is dimensionless.

Some approximate densities at standard conditions:
Air: 0.08 lb/ft^3[1.3 kg/m^3]
Water: 62.4 lb/ft^3[1,000 kg/m^3]
Steel: 462 lb/ft^3[7,400 kg/m^3]
Mercury: 845 lb/ft^3[13,600 kg/m^3]

VISCOSITY

Viscosity is the physical property that characterizes the flow resistance of a fluid. Analogous to thermal conductivity in heat transfer, viscosity may be regarded as momentum conductivity in viscous flow.

For Newtonian fluids, the viscosity is related by:

$$\tau g_c = \mu(du/dy) \qquad\qquad (8\text{-}1)$$

where

$\qquad$ τ = shear stress = momentum flux = Rate of momentum transfer per unit area, lbf/ft^2[N/m^2]
$\qquad$ g_c = 32.17 lb.ft/s^2.lbf[1 kg.m/s^2.N]
$\qquad$ μ = viscosity, lb/ft.s[kg/m.s] = τg_c/(du/dy) = shear stress/shear rate

u = linear velocity, ft/s[m/s]
y = distance perpendicular to direction of flow, ft[m]
du/dy = shear rate, 1/s

The viscosity of Newtonian fluids is independent of time increase and shear rate increase. The viscosity of a Newtonian fluid is sensitive to temperature. For liquids, the viscosity decreases with temperature; however, for gases, it increases according to:

$$\mu = \mu_0(T/273)^n \tag{8-2}$$

Where μ_0 = viscosity at 273 K, T = temperature in K, n varies from 0.65 to 1.
Except for gases at high pressure, viscosity is insensitive to pressure change.
Fluids whose viscosities do not follow the equation(8-1) are called non-Newtonian fluids.

For non-Newtonian fluids, the shear stress and shear rate are related by:

$$\tau g_c = \tau_0 g_c + K'(du/dy)^{n'} \tag{8-3}$$

Where K' is the consistency index, and n' is the rheological constant.

When $\tau_0 > 0$ and n' = 1, then the model represents Bingham plastics. When $\tau_0 = 0$ and n' has a non-zero value other than 1, the model represents non-Newtonian fluids called power law fluids.
When $\tau_0 = 0$ and n' = 1, the model represents Newtonian fluids.

Fluid Type	Effect on viscosity of		n'	Example
	Time increase	Shear rate increase		
Newtonian	No effect	No effect	1	water, air
Non-Newtonian				
Pseudoplastic	No effect	Decrease	0<n'<1	polymer,pigment.
Dilatant	No effect	Increase	>1	Starch slurry, clay slurry,candy.
Thixotropic	Decrease	No effect		asphalt,mayonnaise
Rheopective	Increase	No effect		gypsum slurry

STATIC PRESSURE, STATIC HEAD

$$P_h = h\rho g/g_c = h\gamma \tag{8-4}$$

Where P_h = static pressure, lbf/ft²[N/m²], true on any planet, due to the column of fluid only.
h = head of fluid above the point where static pressure is to be measured = static head, ft[m] of fluid
ρ = fluid density, lb/ft³[kg/m³]

g = 32.17 ft/s^2[9.81m/s^2], variable
g_c = 32.17 lb.ft/s^2.lbf[1 kg.m/s^2.N]
γ = specific weight, lbf/ft^3[N/m^3]

Special cases:

$$P_h = 0.433hs \tag{8-4a}$$

where P_h = static pressure, psi, generally on the surface of the earth, due to the column of fluid only.
  h = static head, ft of fluid.
  s = specific gravity, dimensionless

$$P_h = 9.81hs \tag{8-5}$$

where P_h = static pressure, kN/m^2, generally on the surface of the earth
  h = static head, m of fluid
  s = specific gravity, dimensionless.

Generally, the pressure measured by a gage is relative to the barometric pressure, and is denoted by psi**g**, whereas the absolute pressure is denoted by psi**a**. Thus:

$$\text{Absolute pressure} = \text{Barometric pressure} \pm \text{gage pressure} \tag{8-5a}$$

The minus sign is used for the vacuum gage.

STEADY, INCOMPRESSIBLE FLOW OF FLUID IN CONDUITS AND PIPES

The shear stress at the wall of a pipe, τ_w, due to flow of any fluid, Newtonian or non-Newtonian, is given by:

$$\tau_w = 0.25D(\Delta P/L) \tag{8-6}$$

where ΔP = pressure drop due to skin friction, lbf/ft^2[N/m^2], in a pipe of length L ft[m] and internal diameter D ft[m].

Now, $\Delta P/L$ for laminar flow of Newtonian fluid in a pipe is related to viscosity, velocity or flow rate, and internal diameter as follows:

$$\Delta P/L = 32u\mu/(g_cD^2) \quad \text{[Hagen-Poiseuille Equation]} \tag{8-7}$$

It follows from the two equations mentioned above that the wall shear stress for laminar flow of Newtonian fluid may be represented by:

$$\tau_w = (\mu/g_c)(8u/D) = (\mu/g_c)[32Q/(\pi D^3)] \tag{8-8}$$

where Q is the flow rate ft^3/s[m^3/s]. In general, for laminar or turbulent flow of Newtonian fluids:

$$\tau_w = f\rho u^2/(2g_c) \tag{8-9}$$

For a power-law fluid,
$$\tau_w = 0.25D(\Delta P/L) = K'(8u/D)^{n'} \tag{8-9a}$$

For a power-law fluid, a plot of $0.25D(\Delta P/L)$,lbf/ft²[Pa] vs $8u/D$, s¹ , on a log-log graph would provide the rheological constant, n', the slope, and consistency index, K',lbf/ft²[Pa], at the y-axis corresponding to $8u/D = 1$ at the x-axis.

The generalized equation for frictional pressure drop ΔP,lbf/ft²[Pa] of a power-law fluid for laminar flow may be given by:

$$\Delta P/L = \frac{32K8^{n'-1}u^{n'}}{D^{n'+1}} \tag{8-9b}$$

Where velocity, u is in ft/s[m/s], and diameter, D is in ft[m]. For handling Bingham plastics, and power-law fluids in turbulent flow, see reference[2,4].

The velocity distribution for laminar flow in a circular tube or between planes is given by:

$$u_{local,r} = u_{max}[1 - (r/R)^2] \tag{8-10}$$

where $u_{local,r}$ = local velocity at a distance r, ft/s[m/s]; u_{max} = velocity at the center of the duct or planes, ft/s[m/s]; r = distance from the center line, ft[m]; R = radius of the tube or half the distance between the parallel planes, ft[m].

$$u_{max} = 1.22u \text{ for turbulent flow}(N_{Re} > 10,000)$$
$$= 2u \text{ for circular tubes in laminar flow}$$
$$= 1.5u \text{ for parallel planes, laminar flow, where u = average velocity.}$$

BERNOULLI'S THEOREM

$$Z_A(g/g_c) + u_A^2/(2\alpha g_c) + P_A/\rho_A + W = Z_B(g/g_c) + u_B^2/(2\alpha g_c) + P_B/\rho_B + F \tag{8-11}$$

Where the subscript A denotes upstream conditions, and subscript B denotes downstream conditions, i.e., the flow is from A to B. $\alpha = 1/2$ for laminar flow, 1 for turbulent flow. Z = static head, ft[m], u = velocity, ft/s[m/s], P = pressure, lb/ft²[N/m²], W = work done on the system, ft.lbf/lb[J/kg], F = pressure drop due to skin friction between points A & B, ft.lbf/lb[J/kg], ρ = density, lb/ft³[kg/m³], on earth $g/g_c \sim 1$ lbf/lb[9.81 N/kg]. Each term of the equation represents energy per unit mass of the fluid, ft.lbf/lb[J/kg].

DIMENSIONLESS EQUATION OF REYNOLDS NUMBER

$$N_{re} = Du\rho/\mu = Du/\nu = DG/\mu$$

$N_{Re} < 2100$ indicates viscous(laminar) flow, $N_{Re} > 4000$ indicates turbulent flow, $2100 < N_{Re} < 4000$ indicates transition region. The upper boundary of laminar flow is given by $(N_{Re})_{critical} = 2100$.

Where N_{Re} = Reynolds number, dimensionless; D = diameter of pipe, ft[m]; u = velocity, ft/s[m/s]; ρ = density of fluid, lb/ft^3[kg/m^3]; μ = viscosity of fluid, lb/ft.s[kg/m.s]; ν = kinematic viscosity = μ/ρ, ft^2/s[m^2/s]; G = mass velocity, lb/ft^2.s[kg/m^2.s]

$$1\ cP = 6.72 \times 10^{-4}\ lb/ft.s = 0.001\ kg/m.s$$

DIMENSIONAL EQUATIONS FOR REYNOLDS NUMBER AND PRESSURE DROP

Reynolds Number	**Frictional Pressure Loss [Incompressible Newtonian Fluid]**	
$N_{re} = 123.9du\rho/\mu$	$\Delta P = 0.005176fL\rho u^2/d$	(8-12)
$N_{re} = 50.65Q\rho/(\mu d)$	$\Delta P = 0.000864fL\rho Q^2/d^5$	(8-13)
$N_{re} = 379\rho q/(\mu d)$	$\Delta P = 0.0484fL\rho q^2/d^5$	(8-14)
$N_{re} = 6.32W/(\mu d)$	$\Delta P = 0.0000134fLW^2/(\rho d^5)$	(8-15)

Where:
d = internal diameter, in.; f= Fanning friction factor, dimensionless; L = pipe length, ft.; ΔP = frictional pressure loss, psi; q = flow rate, cuft/min; Q = flow rate, gpm; N_{Re} = Reynolds number, dimensionless; u = velocity, ft/s; W = flow rate, lb/h; μ = viscosity, cP; ρ = density, lb/ft^3.

EQUIVALENT DIAMETER FOR NONCIRCULAR CONDUITS

The equivalent diameter, D_e, is related to the hydraulic radius, r_h, by:

$$D_e = 4r_h$$

where
$$r_h = \frac{\text{cross-sectional area of flow}}{\text{wetted perimeter of channel}}$$

For the annular space of concentric pipes:

$$D_e = D_2 - D_1$$

Where D_2 = inside diameter of outer pipe, and D_1 = outside diameter of inner pipe.

FRICTION FACTOR, NEWTONIAN FLUID

1. For laminar flow($N_{Re} < 2100$), $f = 16/N_{Re}$
2. For turbulent flow

(a) Use Fanning friction factor vs Reynolds number chart, Figure 2-7 in CELR[1],. This may be used for laminar flow also.

(b) For fully turbulent flow, f is independent of Reynolds number:

$$f = \frac{1}{16[\log(3.7D/\epsilon)]^2} \qquad (8\text{-}16)$$

For turbulent flow, f is dependent on Reynolds number:

$$f = \frac{1}{2.444[\ln(0.135\epsilon/D) + 6.5/N_{Re}]^2} \qquad (8\text{-}16a)$$

where D = inside diameter of pipe and ϵ roughness factor of pipe, both in the same unit.

(c) For smooth pipe, such as drawn tubing, glass pipe, etc., and $N_{Re} < 10^5$:

$$f = 0.079 \, N_{re}^{-1/4} \qquad (8\text{-}17)$$

This equation may also be used for turbulent flow of non-Newtonian fluids.

CAUTION! BEFORE USING ANY EQUATION OR FRICTION FACTOR CHART FOR PRESSURE DROP CALCULATION, MAKE SURE THAT THE FRICTION FACTOR IS CONSISTENT WITH THE PRESSURE DROP EQUATION. THE FANNING FRICTION FACTOR USED ABOVE IS ONE-FOURTH OF THE MOODY FRICTION FACTOR.

EQUIVALENT LENGTH OF A PIPING SYSTEM

The total equivalent length of a piping system, L_e, is the sum of straight piping and the equivalent lengths of fittings such as elbows, valves, and other hardware.

$$L_e = L_{straight\ pipe} + \Sigma(L_e)_{fittings} \qquad (8\text{-}18)$$

RESISTANCE COEFFICIENT, K, OF A FITTING

It is the number of velocity heads lost because of the fitting. See CELR[1] table 2.1 for the L_e/D ratio and the resistance coefficient K for various fittings. Multiply L_e/D by the corresponding diameter to get the length of the fitting.

When L_e/D ratio for a specific diameter is not available, it can be estimated from the known L_e/D ratio corresponding to a diameter for fully turbulent flows as follows:

$$(L_e/D)_a = (L_e/D)_b (D_a/D_b)^4 \qquad (8\text{-}19)$$

In the above equations, subscript a refers to the ratio required for a specific diameter D_a, and subscript b refers to known data. It follows from above that the equivalent lengths of pipes of two different diameters for fully turbulent flows are related by:

$$(L_e)_a = (L_e)_b(D_a/D_b)^5 \tag{8-20}$$

When the Reynolds number falls below 1000, the L_e/D ratio should be corrected as follows:

$$(L_e/D)_l = (R_e/1000)(L_e/D)_t \tag{8-21}$$

Where subscript l refers to an L_e/D ratio below 1000, and subscript t refers to L_e/D ratio for turbulent conditions as shown in table 2.1 of CELR.

K and L_e/D are related by:

$$K = 4fL_e/D \tag{8-22}$$

The K factors should be corrected for low velocities as follows:

$$K_a = K_b(f_a/f_t) \tag{8-23}$$

Where K_a refers to the K factor corresponding to friction factor f_a, and K_b is the K factor given in table 2.1 of CELR[1] for friction factor f_t corresponding to fully turbulent conditions.

The frictional pressure drop, K, and L_e/D are related by:

$$\Delta P/\rho = [\Sigma K + 4f\Sigma(L_e/D)](u^2/2g_c) \tag{8-24}$$

Where ΔP is the frictional pressure drop in $lbf/ft^2[N/m^2]$, ρ is the density in $lb/ft^3[kg/m^3]$, L_e & D are the length and diameter respectively of the fitting in the same unit, u is the velocity in ft/s[m/s], g_c = 32.17 $lb.ft/s^2.lbf[$ 1 $kg.m/s^2.N]$

FLOW PAST IMMERSED BODIES, DRAG COEFFICIENT

When a fluid flows past an immersed body, the force in the direction of flow exerted by the fluid on the immersed body is called drag, F_D, $lb_f[N]$. If A_p is the projected area, $ft^2[m^2]$, of the immersed body on a plane perpendicular to the direction of flow, then the drag coefficient, C_D(dimensionless), analogous to the friction factor, f, is given by:

$$F_D/A_p = C_D \rho u^2/2g_c \tag{8-25}$$

Where u is the relative velocity between the immersed body and the fluid. By Newton's third law of motion, an equal and opposite force is exerted by the immersed body on the fluid.
Graphical plots of C_D vs Reynolds number ($D_p u\rho/\mu$) are available in Perry[5th ed. 5-62, 6th ed. 5-64], where D_p is diameter, ft[m], of spherical particle having the same volume as the particle.

COMPRESSIBLE FLOW

A fluid is compressible when the density changes with pressure.

Mach number, N_{Ma} = u/a, where u= velocity of fluid, and a = velocity of sound in the same fluid under identical pressure and temperature, both velocities being in the same unit. $N_{Ma}>1$ indicates supersonic flow. N_{Ma} = 1 indicates sonic flow. $N_{Ma}<1$ indicates subsonic flow.

Velocity of sound, $a = (kg_c RT/M)^{1/2}$ for an ideal gas, $k = C_p/C_v$,   (8-25a)
where a = velocity of sound in an ideal gas, ft/s[m/s]
   C_p = specific heat at constant pressure
   C_v = specific heat at constant volume
   g_c = 32.2 lb.ft/s².lbf [1kg.m/s².N]
   R = 1545 ft.lbf/lb.mol.°R[8314 N.m/kmol.K]
   M = molecular weight of gas,lb/lbmol[g/mol].

Critical pressure ratio. For a constant inlet pressure, the flow rate of a compressible fluid increases as the downstream pressure is lowered until a downstream pressure is reached when the flow rate reaches maximum, and the fluid velocity reaches sonic velocity. At this condition, the ratio of the downstream pressure to upstream pressure in absolute scale is called the **critical pressure ratio.** For an ideal gas:

$$P_{downstream}/P_{upstream} = [2/(k+1)]^{k/(k+1)} \qquad (8-26)$$

If downstream pressure is less than what is calculated from above equation, the flow rate becomes independent of the downstream pressure. As a rule of thumb, when the above ratio is less than 0.5, critical flow(choked flow) is anticipated.

Estimation of flow of compressible fluid.

1. When the pressure drop is < 10% of inlet pressure, or at a very high pressure, use Bernoulli's equation using either inlet or outlet density.
2. When the pressure drop is >10% but <40% of inlet pressure, use Bernoulli's equation, but use the average density based on inlet and outlet conditions.
3. When the pressure drop is >40% of inlet pressure, or for any compressible flow the following method may be used under adiabatic conditions for known upstream pressure P_1, lbf/ft²[N/m²]; upstream temperature, T_1, °R[K]; equivalent pipe length, L, ft[m]; inside diameter of pipe, D, ft[m]; specific heat ratio, k; gas constant, R, 1545, ft.lbf/lbmol.°R [8314 N.m/kmol.K]; molecular weight, M, lb/lbmol[g/mol].

Calculate the friction factor using equation (8-16), and then estimate the upstream Mach number by trial and error using equation (8-27), followed by the calculation of remaining parameters.

$$\frac{(k + 1)}{2} \ln \frac{2+(k + 1)N_{Ma1}^2}{(k + 1)N_{Ma1}^2} - \left(\frac{1}{N_{Ma1}^2} -1 \right) + k(4fL / D) = 0 \qquad (8-27)$$

$$Y_1 = 1 + \frac{(k+1)N_{Ma1}^2}{2} \quad \text{dimensionless} \qquad (8-28)$$

$$T_{choked} = 2Y_1T_1/(k + 1), \,^\circ R[K] \tag{8-29}$$
$$P_{choked} = P_1N_{Ma1}[2Y_1/(k + 1)]^{0.5}, \, lbf/ft^2[N/m^2] \tag{8-30}$$
$$G_{choked} = P_{choked}[kg_cM/(RT_{choked})]^{0.5}, \, lb/ft^2.s[kg/m^2.s] \tag{8-31}$$
$$Mass \, flow = G_{choked}\pi D^2/4, \, lb/s[kg/s] \tag{8-32}$$

TEMPERATURE RISE DUE TO SKIN FRICTION UNDER ADIABATIC CONDITION

Assuming skin friction is completely converted to heat, the temperature rise, ΔT, $^\circ F[K]$, may be given by:

$$\Delta T = \Delta P/(\rho\eta C_p) \tag{8-33}$$

where ΔP is skin frictional pressure drop, $lbf/ft^2[N/m^2]$, ρ is the density, $lb/ft^3[kg/m^3]$
$\eta = 778 \, ft.lbf/Btu[1N.m/J]$, C_p is the specific heat, $Btu/lb.^\circ F[J/kg.K]$.

For the fundamentals of manometers, pump hydraulics, flow meters and compression equipment, see CELR[1].

References:

1. "Chemical Engineering License Review", Das, D. K., and R. K. Prabhudesai, Engineering Press, Austin, TX.

2. A. A. Sultan,"Chemical Engineering", Dec. 19,1988, pp 140-146.

3. G. Goff, Personal communications.

4. "Engineering Flow and Heat Transfer", O. Levenspiel,Plenum Press,NY

Chapter 8
Problems
Transport Phenomenon(Momentum Transfer

Example 8.1 Find the velocity of sound in a gas of density 1 kg/m³ at a pressure of 1.013E5 N/m² with the specific heat ratio of 1.4, assuming the gas is ideal.

Example 8.2 The rheological constant of some power-law fluid, n' is 0.586.
The existing pipe with an internal diameter of 0.154 m is to be replaced with a new pipe of same length to reduce the pressure drop by 50% and maintain the same volumetric flow rate. Assuming the flow is laminar, what size pipe is needed?

Solutions.
Transport Phenomenon.

Solution 8.1.
From equation (8-25a):

$$a = \left(kg_c RT / M\right)^{0.5} = \left(kg_c P / \rho\right)^{0.5}$$

$$= \left(1.4 \times \frac{1kg.m}{s^2 N} \times \frac{1.013E5N}{m^2} \times \frac{m^3}{1kg}\right)^{0.5} = 376.6 \text{ m/s}$$

Solution 8.2.

From equation (8-9b), for the same power-law fluid through the same pipe length:

$$\Delta P \propto u^{n'}/D^{n'+1} \propto Q^{n'}/D^{3n'+1}$$

Therefore, for the same volumetric flow rate Q through the same pipe length:

$$\Delta P_1/\Delta P_2 = 2 = (D_2/D_1)^{2.758} \quad \text{Hence, } D_2 = D_1[10^{(\log 2)/2.758}] = 0.154[10^{0.1091}] = 0.198 \text{ m}$$

Chapter 9

Process Control

Control Systems
Automatic control is required (1) for precise control of the process to produce more uniform and high quality product, (2) for processes which are too rapid for manual control and (3) in hazardous operations where remote control is a necessity. A control system is characterized by an output variable (e.g., temperature) that is automatically controlled through the manipulation of the inputs.

Open control systems:
The inputs to the process are regulated independently without using the measurement of the controlled output variable to readjust the inputs.

A **closed loop or feedback system:**
This implies that the measurement of the controlled variable is used to manipulate one of the process variables.

Feed-forward control:
The measurement of one input variable is used to manipulate another input variable.

Cascade control involves the use of the output of a primary controller to adjust the set point of a secondary controller and is commonly used in feedback control. **The ratio and selector controls** are two other control modes which use two or more interconnected instruments.

Servo-operation : In this mode, the control system is designed to follow the changes in the set point as closely as possible according to some prescribed function.
Regulatory Operation: In regulator operation, the control system is designed to keep output constant i.e. to maintain the controlled variable at a fixed value inspite of the changes in load. This is more common in the control of chemical processes.

On-Off (Bang-Bang) Controller:
In this type, when the measured variable is below the set point, the controller is on with maximum output. However, the controller is off and the output is zero if the measured variable is above the set point. There are several modifications of this controller such as dead band, hysterisis, and differential gap. On-Off control action is inherently cyclic in nature.

Proportional control:
This is also called throttling control. In this control mode , the measurement of the controlled variable is matched against a set point and the error is used to determine the position of the final operator. One drawback of this control is that it can reduce the effect of load change but cannot eliminate it completely giving rise to a steady state difference called offset between the measurement and the set point. Basic equation of the proportional control is as follows

$$p = \frac{100}{P} e + b$$

where P = proportional band, %
     e = fractional change in error or deviation of measurement from the set point
     b = output bias
     p = fractional change in controller output or manipulated variable.

Gain of the controller is the change in its output dp caused by deviation de in the measurement and is expressed as

$$\frac{dp}{de} = K_c G_c$$

The gain of the controller has a steady state component K_c which has a value of 100/P and a dynamic component G_c .Condition for undamped oscillation to persist is,

Gain product of all elements in the loop = $K_c G_c K_p G_P = 1.0$

If the gain product < 1, oscillation will be dampened. Usually a value of 0.5 or less is desirable to avoid exceeding unity which is the limit of stability. Quarter amplitude damping will be achieved by reducing the loop gain to 0.5.

Proportional-Plus-Integral Controller:
To remove the offset of proportional control, an integral action is added to the proportional controller. This integral action integrates the difference between the measurement and the set point and causes the controller's output to change until the error is reduced to zero. Integral action however results in an oscillatory response. Proportional-Plus-Integral controller is represented by the relation

$$p = \frac{100}{P}\left(e + \frac{1}{\tau_i} \int e\, dt\right) \quad \text{where } \tau_i = \text{integral time.}$$

Proportional-Plus-Derivative Control: Derivative action is used to increase the rate of correction. This controller is mathematically represented by the equation

$$p = \frac{100}{P}\left(e + \tau_D \frac{de}{dt}\right) \quad \text{where } \tau_D = \text{ derivative time constant.}$$

Block Diagrams

Block diagrams are used to show functional relationship between the various parts of the control system. The typical components of a block diagram are (a) transfer functions and (b) comparators. These are shown in Figure 9-1. They are obtained directly from the physical system which is divided into sections whose inputs and outputs are distinctly identifiable.

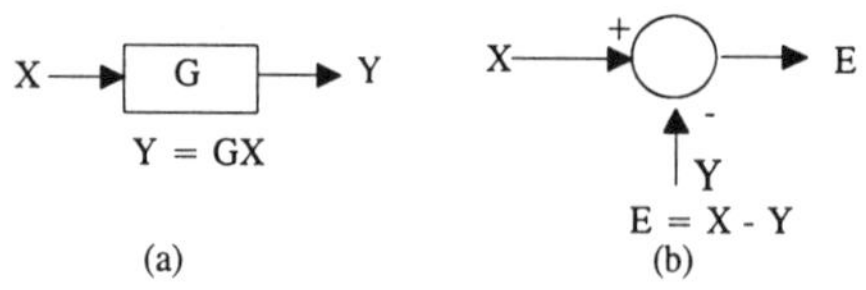

Figure 9-1: Components of a control system block diagram
(a) transfer function (dynamic relationship) (b) comparison of signals

Transfer Functions:
Control system analysis involves the determination of the transfer functions by the application of either mass, energy, or force balance to each part of the system. Developed linear differential equations are solved using Laplace transforms[1,2,3] . A transfer function is defined as follows

$$\textbf{Transfer function} = \frac{\textit{Laplace Transform of Output}}{\textit{Laplace Transform of Input}} = K_c\, G(s)$$

where K_c is the constant portion of the function and is termed the **steady state gain of the controller** while G(s) is the dynamic portion. "(s)" is usually omitted from the transfer function notation in the block diagram. Each block or element of the control system has its own characteristic transfer function.

An overall transfer function of a control system can be obtained by combining the transfer functions of the individual elements with the following rules:
1. For several transfer functions in series, the overall transfer function is the product of the individual transfer functions
2. In a single loop feedback system, the overall transfer function relating any two variables Y and X is given by the equation

$$\text{Overall transfer function} = \frac{Y}{X} = \frac{\pi_f}{1 \pm \pi_l}$$

where π_f = product of transfer functions between locations of X and Y
π_l product of transfer functions in the loop.

Plus sign is to be used in the denominator when the feedback is negative and the minus sign when the feedback is positive. For an example of derivation of the differential equation and its Laplace transform, refer to problem 9-1. Transfer functions for various control actions are given in Table 9-1.

Table 9-1: Transfer Functions of Control Actions

Controller Action	Functional Relationship between Controller Output and Error	Transfer Function
Proportional	$p = K_c e + b$	$\frac{P}{E} = K_c$
Integral	$p = \frac{1}{\tau_i} \int_0^t e\,dt$	$\frac{P}{E} = K_c\left(\frac{1}{\tau_i s}\right)$
Proportional + Integral	$p = K_c\left(e + \frac{1}{\tau_i} \int_0^t e\,dt\right)$	$\frac{P}{E} = K_c\left(1 + \frac{1}{\tau_i s}\right)$
Proportional + derivative	$p = K_c\left(e + \tau_D \frac{de}{dt}\right)$	$\frac{P}{E} = K_c\left(1 + \tau_D s\right)$
Proportional + Integral + derivative	$p = K_c\left(e + \frac{1}{\tau_i} \int_0^t e\,dt + \tau_D \frac{de}{dt}\right)$	$\frac{P}{E} = K_c\left(1 + \frac{1}{\tau_i s} + \tau_D s\right)$

Where p = fractional change in controller output   e = fractional change in error
K_c = controller gain τ_i integral time τ_D derivative time

Proportional Band
Basic relation for a proportional controller is

$$p = K_c e + b$$

where b = bias which is usually adjusted to place the valve in its 50% open position with zero error. The output of the controller equals the bias when there is no error.

$$\% \text{ proportional band} = \frac{100}{Controller\ Gain}$$

or

$$\text{Controller gain} = \frac{100}{\%\ Pr\,oportional\ Band}$$

Proportional band is defined as the percentage of the maximum range of the input variable required to cause 100 % change in the controller output

$$\% \text{ proportional band} = 100 \ \frac{\Delta e}{\Delta e_{max}}$$

where Δe_{max} is the maximum range of the error signal and Δe is the change in error that gives a maximum change in the output of the controller. Substituting the value of K_c into the basic equation for the proportional controller, the following equation results

$$p = \frac{100}{\% \, proportional \, band} e + 50\%$$

The proportional band is also called the proportional band width and is often expressed as the percentage of band width. The amount of the offset can be calculated using the following equation

$$\Delta e = \frac{\% \, proportional \, band}{100} \Delta M$$

where Δe = change in offset and ΔM change in measurement required by load upset.
For a proportional pneumatic controller, the maximum output is 3 to 15 psig (in SI system 20 to 100 kPa) or $\Delta P_{max} = 12 \, psi$ (in SI system, 80 kPa) which is related to error signal by the equation
$$\Delta P_{max} = K_c \Delta e$$

combining definition of % proportional band with the relation for ΔP_{ma} gives

$$\% \, proportional \, band = \frac{100}{K_c} \frac{\Delta P_{max}}{\Delta e_{max}}$$

Control action can be expressed by proportional bandwidth w. This is the error needed to cause a 100 % change in controller output and is usually expressed in terms of chart width. the bandwidth w is given by

$$w = \frac{1}{K_c} \times 100$$

Some pneumatic controllers are calibrated in the sensitivity units. With 4" chart width and 3 to 15 psig output range , the sensitivity of the controller is related to gain by

$$S = 3K_c \; psi/in \; travel$$

Transient Response of Control Systems
The transient response of a control system is obtained with the application of some specified inputs. More commonly used forcing functions with their Laplace transforms are given in Table 9-2.

Table 9-2: Forcing Functions and Their Transforms

Input	Function, f(t)	Laplace Transform
Step	$aU(t)$	a/s
Ramp	at	a/s^2
Impulse	$a[\delta(t)]$	a
Sinusoidal	$\sin \omega t$	$\omega/(s^2 + \omega^2)$

An important consideration in the closed loop control system is the stability of the system. If the response function y(t) in the time domain is bounded as $t \rightarrow \infty$ for all bounded inputs, the system is said to be stable; if $y(t) \rightarrow \infty$, the system is said to be unstable.

First and Second-Order Systems
If the transfer function of a control system has a first order denominator, the system is said to be of first order. Only one parameter, the time constant τ describes the dynamic behavior of such a system. Second

order system requires two parameters to describe its behavior. The denominator of the transfer function of a second order system is called **characteristic equation**. This can be expressed in the quadratic form as

$$\tau^2 s^2 + 2\xi\tau s + 1 = 0 \qquad \text{or} \qquad s^2 + 2\xi\omega s + \omega^2 = 0$$

where
ξ = damping coefficient
ω radian frequency of oscillation $= \sqrt{1-\xi^2}/\tau$
τ period of oscillation

The roots of the above equation will be real or complex depending upon the value of the damping coefficient ξ. The nature of the roots will determine the response function. For three possible types of roots, the system response will be as given in Table 9-3. Special terminology used in describing second order undamped behavior of control systems will now be given.

Table 9-3: Responses of a Second-Order System

ξ	Nature of Roots	Response
< 1	Complex	Oscillatory or underdamped
= 1	Real and equal	Critically damped
> 1	Real	Nonoscillatory or overdamped

Overshoot for a unit step $= \exp\left(-\pi\xi/\sqrt{1-\xi^2}\right)$

Decay ratio $= \exp\left(-2\pi\xi/\sqrt{1-\xi^2}\right)$

Rise time: time required for the response to first reach its ultimate value.

Response time: time required for the response to come within $\pm$certain percent (usually ± 5 %) of its ultimate value.

Period of oscillation : radian frequency, $\omega = \dfrac{\sqrt{1-\xi^2}}{\tau}$ radians/time

Cyclical frequency , $f = \dfrac{1}{T} = \dfrac{1}{2\pi}\dfrac{\sqrt{1-\xi^2}}{\tau}$ also $\omega = 2\pi f$ T = period of oscillation = time/cycle

Natural period of oscillation:

When damping is eliminated ($\xi = 0$), the system oscillates continuously without attenuation in amplitude. The radian frequency under these undamped condition is $1/\tau$ and is referred to as natural frequency ω_n.

$$\omega_n = 1/\tau$$

Natural frequency $= f_n = 1/T_n = 1/2\pi\tau$

$$\frac{Actual\ frequency}{Natural\ frequency} = \frac{f}{f_n} = \sqrt{1-\xi^2}$$

For more advanced topics, such as stability of higher order (>2) systems, Routh's test, frequency analysis, bode diagram, root locus method, and feed -forward control, the reference should be consulted[4].

Problems

9-1: Derive a transfer function for the transient variation in the liquid height for the tank filling system shown below and prepare a block diagram for the control system.

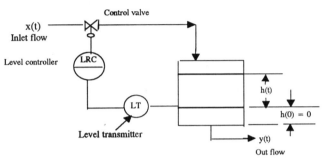

: Sketch of liquid filling system for problem 9-1

9-2: A block diagram of a control system is shown below. Obtain the overall transfer function C(s)/R(s).

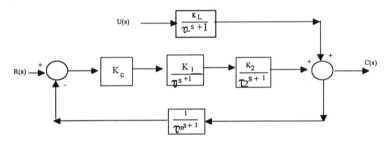

9-3 : The liquid level in the bottom of a distillation column is controlled with a pneumatic proportional controller by throttling a control valve located in the bottom discharge line. The level may vary from 15 cm. to 105 cm. from the bottom tangent line of the column. With the controller set point held constant, the output pressure of the controller varies from 100 kPag (valve fully closed) to 20 kPag (valve fully open) as the level increases from 30 to 90 cm. from the bottom tangent line.

    (a) Find the % proportional band and the gain of the controller

    (b) If the proportional band is changed to 80 % , find the gain and the change in level required to cause the valve to go from fully open to a fully closed position.

Solutions to Practice Problems

9-1:

A material balance, over a difference time element Δ with assumption of constant density ρ can be written as follows

$$\rho A h_2 - \rho A h_1 = \rho \, \overline{x(t)} \, \Delta t - \overline{\rho y(t)} \, \Delta t$$

Which can be written in difference form after cancellation of ρ as follows

$$A \frac{\Delta h(t)}{\Delta t} = \overline{x(t)} - \overline{y(t)}$$

Taking limit as t → 0 gives

$$A \frac{dh(t)}{dt} = x(t) - y(t)$$

which on rearrangement becomes

$$x(t) = y(t) + A\frac{dh(t)}{dt}$$

where A is the cross sectional area of the tank. Laplace transformation gives

$$X(s) = Y(s) + A[sH(s) - h(0)]$$

For simplification , assume h(0) = 0. Therefore

$$X(s) = Y(s) + AsH(s)$$

Then the transfer function for the level system is given by

$$\frac{Laplace\ transform\ of\ output}{Laplace\ transform\ of\ input} = \frac{H(s)}{X(s) - Y(s)} = \frac{1}{As} = K_L G_L(s)$$

1/As is the transfer function of the filling system. It comprises of a constant portion $K_L = 1/A$ and $G_L(s) = 1/s$ which is the dynamic portion. Figure 9-2 shows the block diagram for the liquid filling system. The transfer function for the filling system is shown as $K_L G_L$

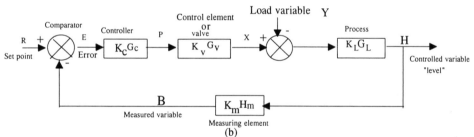

Figure 9-2 : Block diagram for liquid filling system

9-2:

For transfer function between C and R ,

$$\pi_f = K_c \frac{K_1}{\tau_1 s + 1} \frac{K_2}{\tau_2 s + 1}$$

$$\pi_l = K_c \left[\frac{K_1}{\tau_1 s + 1} \right] \left[\frac{K_2}{\tau_2 s + 1} \right] \left[\frac{1}{\tau_m s + 1} \right]$$

Then

$$\frac{C}{R} = \frac{K_c \left[K_1/(\tau_1 s + 1) \right] \left[K_2/(\tau_2 s + 1) \right]}{1 + K_c K_1 K_2 / \left[(\tau_1 s + 1)(\tau_2 s + 1)(\tau_m s + 1) \right]}$$

Which simplifies to

$$\frac{C}{R} = \frac{K_c K_1 K_2 (\tau_m s + 1)}{(\tau_1 s + 1)(\tau_2 S + 1)(\tau_m s + 1) + K_c K_1 K_2}$$

9-3 :

(a) calculation of the proportional band

$$\text{proportional band} = 100\frac{\Delta e}{\Delta e_{\max}}$$

$$= \frac{span\ of\ contr\ variable\ for\ fully\ closed\ position\ of\ control\ valve}{full\ span\ or\ range}$$

$$= \frac{0.9-0.3}{1.05-0.15} = 67\%$$

$$\text{Gain of the controller} = \frac{\Delta P}{\Delta e} = \frac{100-20}{0.9-0.3} = 133.3 \text{ kPa/m}$$

(b) If the proportional band is 80 %, $\frac{\Delta e}{\Delta e_{\max}} = 0.\ (1.05 - 0.15) = 0.72$ m

$$\text{Gain of controller} \quad K_c = \frac{\Delta p}{\Delta e} = \frac{100-20}{0.72} = 11 \text{ kPa/m}$$

References:

1. Das D.K. and R. K. Prabhudesai, **Chemical Engineering License Review**; Engineering Press, 2d ed. San Jose, CA (1996).
2. Wylie R., **Advanced Engineering Mathematics,** 4th ed., McGraw-Hill, New York, 1975 pp 268
3. Perry R.H. and D. Green, **Handbook for Chemical Engineers,** McGraw-Hill, New York, 6th ed.. (1983)
4. AIChE Modular Instruction series, **Process Control,** vol 1-4 (1987)

Chapter-10
PROCESS EQUIPMENT DESIGN
FLUID HANDLING EQUIPMENT

Centrifugal Pumps

1. Relation between volumetric flow rate, Q, and mass flow rate, W

$$Q = W/(500s) \qquad (10\text{-}1)$$

2. Net Positive Suction Head Available(NPSH$_A$)

$$NPSH_A = (P_1 - P_v - \Delta P_{f1})2.31/s + Z_1 + u^2/(2g_c) \qquad (10\text{-}2a)$$

When NPSH is determined from the gage reading at the pump suction:

$$NPSH_A = (P_1 \pm R_G - P_v - \Delta P_{f1})2.31/s + u^2/(2g_c) + Z_{1p} \qquad (10\text{-}2b)$$

NPSH$_A$ must always be positive, and greater than NPSH$_{required}$.

3. Suction Pressure, P$_s$

$$P_s = P_1 + 0.433Z_1 s - \Delta p_{f1} \qquad (10\text{-}3)$$

4. Discharge Pressure, P$_d$

$$P_d = P_2 + \Delta P_{f2} + 0.433Z_2 s \qquad (10\text{-}4)$$

5. Total dynamic head, TDH.

TDH is independent of fluid density, although horsepower, discharge pressure etc. would change with density.

$$\begin{aligned} TDH &= (P_d - P_s)2.31/s \\ &= (P_2 - P_1 + \Delta P_{f1} + \Delta P_{f2})2.31/s + Z_2 - Z_1 \end{aligned} \qquad (10\text{-}5)$$

6. Hydraulic horsepower, hhp

$$hhp = \frac{W(TDH)}{1.98 \times 10^6} = \frac{Q(TDH)s}{3960} = \frac{Q(P_d - P_s)}{1714} \qquad (10\text{-}6)$$

7. Brake horsepower, bhp

$$bhp = hhp/\varepsilon \qquad (10\text{-}7)$$

8. Motor horsepower, mhp

$$mhp = bhp/\varepsilon_m \qquad (10\text{-}8)$$

9. Maximum temperature rise due to pumping, assuming all power input goes to heat the fluid, ΔT

$$\Delta T = TDH/(778C_p\varepsilon_{overall}) \qquad (10\text{-}9a)$$
$$dT/d\theta = 5.1(bhp_{shut\text{-}off})/(VsC_p) \qquad (10\text{-}9b)$$

10. Minimum recirculation flow.

When a running pump is put on a standby mode, such as in a batch operation, a minimum flow, Q_{min}, is to be drawn from the pump back to the suction vessel, normally through a restriction orifice, for thermal protection of the pump.

$$Q_{min} = \frac{5.1(bhp_{shut\text{-}off})}{s[C_p \cdot \Delta T_{safe} + 0.001285(TDH_{shut\text{-}off})]} \qquad (10\text{-}10)$$

ΔT_{safe} = Saturation temperature corresponding to $[P_v + (0.433s)NPSH]$ - operating temperature, $^\circ F$

This minimum flow should be added to the maximum demand to estimate the design capacity of a batch pump expected to run on a standby mode.
There is still another minimum flow called the mechanical minimum flow which should be maintained to avoid excessive vibration and shaft deflection due to unbalanced radial loads. This flow is supplied by the manufacturer of the pump. The final minimum flow should be greater of the two.

11. Pump Affinity Laws

	WITHIN THE SAME PUMP		**GEOMETRICALLY SIMILAR**	
	D constant	**N constant**	**PUMPS, N constant**	
1. Capacity	$Q_2/Q_1 = N_2/N_1$	$Q_2/Q_1 = D_2/D_1$	$Q_2/Q_1 = (D2/D1)^3$	(10-11)
2. Head	$H_2/H_1 = (N_2/N_1)^2$	$H_2/H_1 = (D_2/D_1)^2$	$H_2/H_1 = (D_2/D_1)^2$	(10-12)
3. Power	$bhp_2/bhp_1 = (N_2/N_1)^3$	$bhp_2/bhp_1 = (D_2/D_1)^3$	$bhp_2/bhp_1 = (D_2/D_1)^5$	(10-13)

The same affinity laws apply to centrifugal fans and blowers. For design information of fans, and compressors see reference[1].

where:
bhp = Brake horsepower, $bhp_{shut\text{-}off}$ = brake horsepower at shut-off
C_p = Specific heat of fluid pumped, Btu/lb.$^\circ F$
D = Impeller diameter, ft
$dT/d\theta$ = Rate of temperature rise in a dead-headed pump, $^\circ F$/minute
g_c = Newton's law proportionality factor, 32.17 lb.ft/s^2.lbf
H = Total dynamic head, ft
hhp = hydraulic horsepower
mhp = motor horsepower
N = impeller speed, revolutions per minute(rpm)

$NPSH_A$ = Net Positive Suction Head available, ft

P_1 = Pressure in suction vessel, psia

P_2 = Pressure in discharge vessel, psia

ΔP_{f1} = Total frictional drop in suction line, psi

ΔP_{f2} = Total frictional drop in discharge line, psi

P_d = Pump discharge pressure, psia

P_s = Pump suction pressure, psia

P_v = Vapor pressure of fluid, psia, at the operating temperature

Q = volumetric flow rate, gpm

Q_{min} = Minimum flow rate of pump for thermal protection, gpm

R_G = Gage reading, psig. Use minus sign for vacuum gage.

s = specific gravity of fluid being pumped, dimensionless

TDH = Total dynamic head, ft

$TDH_{shut-off}$ = Total dynamic head at shut-off,ft

ΔT = Maximum temperature rise due to pumping, °F

u = Velocity of fluid in suction line ,ft/s

V = holding capacity of the casing of the pump, gallon

W = mass flow rate, lb/h

Z_1 = Vertical distance between the liquid surface in the suction vessel and the centerline of the pump, ft. A negative value of Z_1 must be used in equations (10-2a), (10-3), and (10-5), if the liquid surface is below the centerline of the pump.Z_{1p} =distance between the centerline of the gage and the center line of the pump,ft. Use minus sign if the centerline of the gage is below the centerline of the pump.

Z_2 = Vertical distance between the liquid surface in the discharge vessel for subsurface discharge or the point of discharge for above-surface discharge and the centerline of the pump,ft

ε = Pump efficiency, decimal

ε_m = Motor efficiency, decimal

$\varepsilon_{overall} = \varepsilon\varepsilon_m$

Problem 10.1 A centrifugal pump equipped with a motor having 7.5 KW (brake) at shut-off head and holding 0.006 m³ of fluid of specific gravity 1.1 and specific heat of 4186 J/kgK in the casing is running dead-headed. Find the rate of temperature rise.

Problem 10.2 Determine the minimum flow for thermal protection of a pump delivering water at 122°F with a shut-off head of 1032 ft and shut-off bhp of 48. The available NPSH is 12 ft.

Agitator

1. The agitator behaves like a pump. The primary pumping capacity, Q, is given by:

$$Q = 4.33 \times 10^{-3} N_Q N D^3 \tag{10-14}$$

2. The apparent superficial velocity, v_b, due to pumping effect of the impeller is given by:

$$v_b = Q/(A \times 7.48) \tag{10-15}$$

3. Chemscale, N_I. This is an arbitrary scale of agitation(1= mild, 10=violent).

$$N_I = v_b/6 \tag{10-16}$$

v_b, expressed in ft/minute, varies from 6(mild) to 60(violent).

4. Tip speed: the linear velocity of the tip of the impeller blade

$$TS = 0.2618DN \qquad (10\text{-}17)$$

There is an arbitrary scale of agitation on the basis of tip speed. Low(TS= 200-500), Medium(TS= 500-900), High(900-2000)

5. Turnover time: time to circulate the entire contents of the vessel once.

$$\theta_Q = V/Q \qquad (10\text{-}18)$$

6. Mixing time[3], θ: time to mix materials to satisfy a specified fluctuation of concentration of a key ingredient(e.g. $\pm$ 0.1%).

For conventional axial flow impellers(Reynolds number$>1\times10^5$):

$$(N/K)(D/T)^2 = N_P = 0.9 \qquad (10\text{-}19a)$$

$$a = 2e^{-K\theta} \qquad (10\text{-}19b)$$

$$\theta = 0.9\ln(2/a)(T/D)^2/N \qquad (10\text{-}19c)$$

With $a = 0.001$,

$$\theta = 6.84(T/D)^2/N \qquad (10\text{-}19d)$$

7. Reynolds Number, N_{Re}

$$N_{Re} = 10.7sND^2/\mu \qquad (10\text{-}20)$$

8. Power Number, N_P

$$N_P = \frac{1.523\times10^{13}P}{sN^3D^5} \qquad (10\text{-}21)$$

9. Torque, τ

$$\tau = 63025P/N \qquad (10\text{-}22)$$

10. Motor Load (3 phase power)

$$HP = \frac{1.73 \, (\text{Voltage})(\text{ampere})(\%\text{motor efficiency})(\%\text{power factor})}{746}$$ (10-23)

Where:
a = amplitude of concentration variation, decimal
A = vessel cross section, ft^2
D = impeller diameter, inch
K = amplitude decay constant, min^{-1}
N = rpm of impeller
N_I = chemscale
N_Q = pumping number(dimensionless). It depends on the type of impeller and impeller Reynolds number, and is given by the manufacturer of agitator.
N_P = power number, dimensionless, dependent on impeller design and Reynolds number.
P = agitator power, HP
Q = primary pumping capacity, gpm
s = specific gravity of fluid, dimensionless
T = Tank diameter, inch
TS = tip speed, ft/minute
v_b = apparent superficial velocity, ft/min
V = working capacity of the vessel, gallon
θ = mixing time, minute
θ_Q = turnover time, minute
τ = torque, inch-lb
μ = viscosity, cP

Problem 10.3 A small amount of hydrochloric acid is added for pH control to a baffled reactor, 2.74 m inside diameter, fitted with a 1.92 m axial flow impeller. Estimate the rps and power of the agitator if the specific gravity is 1.15, and mixing time of 10 seconds with ± 0.1% concentration fluctuation is required to avoid side reactions.

Heat Transfer Equipment

For the estimation of heat transfer coefficient , please see reference[1].

Pressure Vessels

The design and construction of pressure vessels are governed by the codes of the American Society of Mechanical Engineers(ASME) and the standards of the American Petroleum Institute(API). Three sections of ASME that governs the pressure vessels are:

(1) ASME SECTION I for fired pressure vessels such as boilers.

(2) ASME SECTION VIII Division 1 and 2 for unfired metallic pressure vessels.
SECTION VIII Division I is used to design most of the pressure vessels required by the chemical industries. It employs relatively approximate formulas to specify a vessel whose design pressure(internal or external) is greater than 15 psig but does not exceed 3000 psig with an internal diameter exceeding 6 inches. Division 2 uses more complex and rigorous mathematical analysis, and is used for application involving severe service(e.g. toxic chemicals), cyclic operations, design pressure exceeding 3000 psig, etc.
(3) ASME SECTION X for fiber-reinforced thermosetting plastic pressure vessels for design pressure exceeding 15 psig but not exceeding 3000 psig, temperature range of -65 °F to 250 °F for non-lethal service. External pressure must not exceed 15 psig.
API standards (12D, 12F, 620, 650, 650F) cover design of pressure vessels with design pressure not exceeding 15 psig.

Design for internal pressure per ASME SECTION VIII Division 1

Maximum Allowable Working Pressure(MAWP)

It is the least of the pressure ratings of the various components of a pressure vessels. The pressure ratings of the components are back-calculated by using the code formulae from known plate thickness after allowances are made for tolerences, corrosion, thinning due to fabrication, and other loadings. The MAWP at hot and corroded condition is valid only when the vessel is hydrotested at 1.5 times the MAWP at new and cold condition (because of higher thickness), and stamped along with the design temperature on a plate attached to the vessel.

Design Pressure

This pressure, taken at the highest connection of the pressure vessel, is addded to the hydrostatic head, and the resulting pressure is used to calculate the theoretical thickness of a plate.

Code Formulas for Calculating Plate Thickness and Maximum Allowable Working Pressure

	Wall thickness inch	Maximum allowable working pressure,psig	
Cylindrical shell:	$t = \dfrac{PR}{SE + 0.4P}$	$P = \dfrac{SEt}{R - 0.4t}$	(10-24)
Sphere and hemispherical head:	$t = \dfrac{PR}{2SE + 0.8P}$	$P = \dfrac{2SEt}{R - 0.8t}$	(10-25)

2:1 Ellipsoidal head:
$$t = \frac{PR}{SE + 0.9P} \qquad P = \frac{SEt}{R - 0.9t} \qquad (10\text{-}26)$$

Conical head:
$$t = \frac{PR}{\cos\alpha(SE + 0.4P)} \qquad P = \frac{SEt(\cos\alpha)}{R - 0.4t(\cos\alpha)} \qquad (10\text{-}27)$$

ASME flanged & dished head(torispherical): (L/r = 16 2/3)[1]
$$t = \frac{0.885PL}{SE + 0.8P} \qquad P = \frac{SEt}{0.885L - 0.8t} \qquad (10\text{-}28)$$

Where:

E = joint efficiency, decimal, varies from 0.45 to 1.0

L = outside radius of dish for ASME flanged & dished head, inch

P = design pressure or maximum allowable working pressure, psig

R = outside radius of vessel, inch

r = knuckle radius of a flanged & dished head

S = allowable stress of material corresponding to the design temperature, psi

t = wall thickness, inch. Corrosion allowance, if applicable, and other appropriate allowances are added to this thickness, and the nearest commercially available plate thickness is chosen.

α = one-half of the included apex angle(cone angle) of a conical head

Notes:

1. For L/r < 16 2/3 see ASME Sectin VIII

2. For external pressure see ASME Section VIII

Problem 10.4 Calculate the minimum thickness required for a cylindrical shell with outside radius of 54 inches. The shell is constructed of carbon steel with allowable stress of 17,500 psi. The joint efficiency is 85%, and the corrosion allowance is 0.125 inch. The design pressure is 100 psig, with no hydrostatic head.

Mass Transfer columns

Mass transfer columns are broadly divided into packed columns and plate columns. The companion book[1] outlines the method of sizing packed columns. Sizing of plate columns involves selection of tray spacing, and sizing of the diameter and overall height. Plate spacing should be sufficient for the seperation of the entrained liquid from the vapor before it reaches the next higher plate. Plate spacing is a function of many variables. It generally varies from 12 inches to 48 inches. Flooding limits for bubble-cap and perforated plates are plotted in Fig 18-10 of Perry[4]. For a specified tray spacing, and known vpor and liquid flow rates, known physical properties such as liquid density, vapor density, and liquid surface tension, the flooding velocity can be calculated. Design velocity ranging from 80% to 95% of the flooding velocity is used to estimate the net column area from a known vapor load. Before selecting the design velocity, check Fig. 18-22 of reference[4] so that entrainment is less than 0.1 lb of liquid per pound of liquid flow. Depending on the chosen tray type, an area required for the downcomer is to be added to get the total cross section area of the column. Liquid loading in gpm and a recommended liquid velocity (gpm/ft^2) are used to estimate the downcomer area.

Determination of actual number of plates (N_a) requires the knowledge of overall plate efficiency(E_{oc}) and the number of theoretical plates(N_t):

$$E_{oc} = \frac{N_t}{N_a}$$

The overall plate efficiency can be determined by empirical methods such as O'Connell correlation, Fig 18-23a in Perry[4], Bakowski correlation [equation (18-29)] in Perty or by the theoretical predictive method of AICHE as shown in Perry[4](pp 18-15).

Example 10-5. Size the diameter of a column, using Fair method, for bubble-cap trays, at 80% of the flooding velocity with 24 inches of tray spacing. Data available are:

    Vapor flow : 300,000 lb/h
    Liquid flow: 284,600 lb/h
    Vapor density: 3 lb/ft^3
    Liquid density: 30 lb/ft^3
    Surface tension: 20 dyn/cm
    Recommended downcomer velocity: 100 gpm/ft^2
(A)4 ft   (B) 16 ft   (C) 12 ft   (D) 9 ft

Solutions

Problem 10.1

0.0006 m^3 x 1.1 x 1000 kg/m^3 x 4186 J/kgK x dT/dθ = 7.5 x 10^3
    dT/dθ = 0.2715 °C/s

Problem 10.2

From equation(10-10):

$$Q_{min} = \frac{5.1(bhp_{shut-off})}{s[C_p.\Delta T_{safe} + 0.001285(TDH_{shut-off})]}$$

ΔT_{safe} = Saturation temperature corresponding to [P_v + (0.433s)NPSH]
    - operating temperature, °F

P_v of water at 122°F   = 1.8 psia
0.433sNPSH = 0.433x.99x12 = 5.1 psia
                              ————
                         6.9 psia

Saturation temperature of water corresponding to 6.9 psia = 176 °F
ΔT_{safe} = 176-122 = 54

$$Q_{min} = \frac{5.1 \times 48}{0.99 \times (1 \times 54 + 0.001285 \times 1032)} = 4.5 \text{ gpm}$$

Problem 10.3
From equation (16)

$$\theta = 0.9\ln(2/a)(T/D)^2/N$$

$$N = 0.9\ln(2/a)(T/D)^2/\theta$$

$$a = 0.001, T/D = 2.74/1.92 = 1.427, \theta = 10 \text{ s}$$

$$N = 1.39 \text{ rps} \simeq 1.4 \text{ rps}$$

From equation(18)

$$N_P = \frac{P_I}{N_{rps}^3 D^5 s} = 0.9$$

$$P_I = N_P N_{rps}^3 D^5 s$$
$$= 0.9(1.4)^3(1.92)^5 \times 1.1$$
$$= 70.9 \text{ KW}$$

Problem 10.4

$$t = \frac{PR}{SE - 0.6P} = \frac{100 \times 54.125}{17500 \times .85 - 0.6 \times 100} = 0.364 \text{ inch}$$

Theoretical minimum thickness = 0.364 + 0.125 = 0.489 inch. Note that corrosion allowance is added to the outer radius.

Problem 10.5

The vapor velocity through the net area of a distillation column may be computed by Fair's method as outlined in Perry[4](pp 18-6 & 18-7):

$$U_{nf} = C_{sb,flood} \left(\frac{\sigma}{20} \right)^{0.2} \left(\frac{\rho_l - \rho_g}{\rho_g} \right)^{0.5}$$

To calculate $C_{sb,flood}$, one has to calculate the flow parameter F_{lv} :

$$F_{lv} = \frac{L}{G} \left(\frac{\rho_g}{\rho_l} \right)^{0.5}$$

$$= (284600/300000)(3/30)^{0.5} = 0.3$$

From Fig. 18-10, $C_{sb,flood} = 0.24$.

The flooding velocity, U_{nf}:

$$U_{nf} = 0.24 \left(\frac{30-3}{3} \right)^{0.5} = 0.72 ft/s$$

The design velocity, $U_{nd} = 0.8 \times 0.72 = 0.576$ ft/s
Net flow area = $(300000/3600 \times 3 \times 0.576) = 48.23$ ft^2.
Liquid flow = $(284600 \times 62.4/500 \times 30) = 1183.94$ gpm.
Downcomer area = $1183.94/100 = 11.84$ ft^2 .
Total area = 60.07 ft^2. Diameter = $(4 \times 60.07/3.1416)^{0.5} = 8.74$ ft.
Answer: D

References:
1. "**Chemical Engineering License Review**", D.K. Das and R. K. Prabhudesai, Engineering Press, Austin, TX
2. "**Chemical Engineering**",February 23, 1981, pp 83-85
3. S. J. Khang and O. Levenspiel, **Chemical Engineering Science**, 1976, Vol 31, pp 5610-577, Pergamon Press, GB.
4. "**Perry's Chemical Engineers' Handbook**", 6th edition, p 18-7, McGraw-Hill Inc., New York.

Chapter 11
Computers and Numerical Methods
Lincoln D. Jones

Introduction

The Fundamentals of Engineering Exam contains seven questions concerning computers in the morning session and three questions in the afternoon session. These questions cover the topics of operating systems, networks, interfaces, spreadsheets, flow charting, and data transmission. Each of the branch specific afternoon exams contain three questions on numerical methods related to that branch.

You should review the glossary of important computer related keywords that is presented at the end of this chapter to ensure you have a basic understanding of this broad general topic. Should you find any term unfamiliar to you, a review of that topic would be warrented. The current exam does not include a programming language such as FORTRAN or BASIC but use of applications such as spreadsheets are included. You should be familiar with one of the popular spreadsheet programs such as Excel, Quattro Pro or 1-2-3.

Number Systems

The number system most familiar to everyone is the decimal system based on the ten symbols 0 through 9. This base 10 system requires ten different digits to create the representation of numbers.

A far simpler system is the binary number system. This base 2 system uses only the characters 0 and 1 to represent any number. A binary representation 110, for example, corresponds to the number
$$1x2^2 + 1x2^1 + 1x2^0 = 4 + 2 + 0 = 6$$
Similarly the binary number 1010 would be
$$1x2^3 + 0x2^2 + 1x2^1 + 0x2^0 = 8 + 0 + 2 + 0 = 10$$

The digital computer is based on the binary system of on/off, yes/no, or 1/0. For this reason it may at times be necessary to convert a decimal number (like 12) to a binary number (for 12 it would be 1100).

Example 1
The binary number 1110 corresponds to what decimal (base 10) number?

Solution
$$1x10^3 + 1x10^2 + 1x10^1 + 0x10^0 = 8 + 4 + 2 = 14$$

Data Storage
Memory chips include random access memory (RAM), read only memory (ROM),and programmable read only memory (PROM).

Diskettes have been the primary way to move data about on personal computers. Initially diskettes were 5-¼ inch "floppies" (360K to 2MB of data) that could be damaged easily. They have since been replaced by 3-½ inch diskettes (720 to several MB of data) in a hard plastic housing.

Hard drives (currently about 500MB to 2GB of data) are units with the hard disks permanently sealed in a module. Other storage devices are removable hard disk cartridges, optical disks, read-only media (ROM), write once, read many media (WORM), and compact disk read-only memory (CD-ROM).

Data Transmission

A computer produces digital signals (represented as on-off or 0-1), but these signals often must be transmitted on voice transmission (analog) systems. The conversion of digital signals to analog signals is called *modulation*. The subsequent conversion back to digital from analog is called *demodulation*. The device doing this conversion is called a *modem* (short for modulation-demodulation). Early modems had speeds of 300-1200 bits per second, but modern modems operate at 14,400-28,800 bits per second.

Data transmission requires that the equipment at both the sending and receiving locations must be able to "talk" to each other. Two methods commonly used are asynchronous and synchronous transmission. In asynchronous transmission a start signal is sent at the beginning and a stop signal at the end of each character. Synchronous transmission, on the other hand, is where a bit pattern is transmitted at the beginning of the message to synchronize the internal clocks of the sending and receiving devices.

Programming

Computer programming may be thought of as a four-step process:
1. Defining the problem
2. Planning the solution
3. Preparing the program
4. Testing and documenting the program

Once the problem has been carefully defined, the basic programming work of planning the computer solution and preparing the detailed program can proceed. In this section the discussion will be limited to two ways to plan a computer program to solve a problem:
    Algorithmic Flowcharts
    Pseudocode

Algorithmic Flowcharts

An algorithmic flowchart is a pictorial representation of the step-by-step solution of a problem using standard symbols. Some of the commonly used figures are shown in Figure 11-1. Consider the following simple problem.

Example 2 ————————————————————————————————

A present sum of money (P) at an annual interest rate (I), if kept in a bank for N years would amount to a future sum (F) at the end of that time according to the equation $F = P(1+I)^N$. Prepare a flowchart for $P = \$100$ $I = 0.07$ $N = 5$ years and compute and output the values of F for all values of N from 1 to 5. Figure 11-2 is a flowchart for this situation.

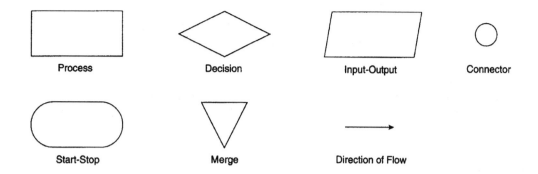

Figure 11-1 Flowchart Symbols

Figure 11-2

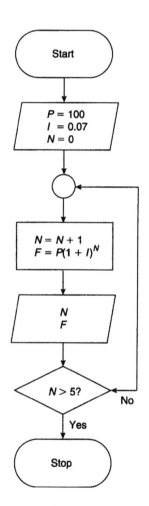

Example 3 ────────────────────────────────────
Consider the flowchart in Figure 11-3.

Figure 11-3

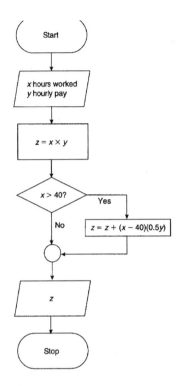

The computation does which of the following?
    (a) Inputs hours worked and hourly pay and outputs the weekly paycheck for 40 hours or less.
    (b) Inputs hours worked and hourly pay and outputs the weekly paycheck for all hours worked including over 40 hours at premium pay.

The answer is (b).

Pseudocode
    Pseudocode is an English-like language representation of computer programming. It is carefully organized to be more precise than a simple statement, but may lack the detailed precision of a flowchart or of the computer program itself.

Example 4
Prepare pseudocode for the computer problem described in example 11-2.
Solution
INPUT P,I, and N = 0
DOWHILE   N < 5
COMPUTE N = N + 1
        $F = P (1 + I)^N$
OUTPUT   N,F
IF N > 5 THEN ENDO

Spreadsheets

For today's engineers the ability to create and use spreadsheets is essential. The most popular spreadsheet programs are Microsoft's **Excel**, Novell's **Quattro Pro**, and Lotus **1-2-3**. Each of these programs use similar construction, methods, operators, and relative references.

Three type of information may be entered into a spreadsheet: text, values, or formulas. Text includes labels, headings and explanatory text. Values are numbers, times or dates. Formulas combine operators and values in an algebraic expression.

A **cell** is the intercept of a column and a row. Its location is based upon its column-row location, for example B3 would be the intercept of column B and row 3. Column labels are across the spreadsheet and row labels are on the side. To change a cell entry, the cell must be highlighted using either an address or pointer.

A group of cells may be called out by using a **range**. Cells A1, A2, A3, A4 could be called out using the range reference A1:A4 (or A1..A4). Similarly, A2,B2,C2,D2 would use the range reference A2:D2 (or A2..D2).
In order to call out a block of cells, a range call out might be A2:C4 (or A2..C4) and would reference the following cells:
A2 B2 C2
A3 B3 C3
A4 B4 C4

Formulas may include cell references, operators (such as +,-,*,/) and functions (such as SUM, AVG). The following formula   SUM(A2:A6) or SUM(A2..A6) would be evaluated as equal to A2+A3+A4+A5+A6.

Relational References
Most spreadsheet references are relative to the cell's position. For example if the content of cell A5 contained B4 then the value of A5 is the value of the cell up one and over one. The relational reference is most frequently used in table tabulations as the following example for an inventory where cost times quantity equals value and the sum of the values yields the total inventory cost.

Inventory Valuation

	A	B	C
1 Item	Cost	Quantity	Value
2 box	5.2	2	10.4
3 tie	3.4	3	10.2
4 shoe	2.4	2	4.8
5 hat	1.0	1	1.0
6 Sum			26.4

In C2 the formula A2*B2
In C3 the formula A3*B3 and so forth.
For the Summation use the function SUM. In C6 use SUM(C2:C5)

Instead of typing in each cell's formula, the formula can be copied from the first cell to all of the subsequent cells by first highlighting cell C2 then dragging the mouse to include cell C5. The first active cell is C2 would be displayed in the edit window. Type the formula for C2 as A2*B2 and hold the control key down and the enter key is pressed. This copies the relational formula to each of the highlighted cells. Since the call is relational, then in cell C3 the formula is evaluated as A3*B3. In C4 the cell is evaluated as A4*B4. Similarly, edit operations to include copy, or fill operations simplify the duplication of relational formula from previously filled in cells.

Arithmetic Order of Operation
Operations in equations use the following sequence: exponentiation, multiplication and division, followed by addition and subtraction. Parentheses in formulas override normal operator order.

Absolute References
Sometimes a reference to a cell must be used that should not be changed, such as an data variable. An absolute reference can be named by inserting a dollar sign "$" before the column-row reference. If B2 is the data entry cell, then by using B2 as its reference in another cell, the call will always be evaluated to cell B2, even though it may be copied to another cell. Mixed reference can be made by using the dollar sign for only one of the elements of the reference. The reference B$2 is a mixed reference in that the row does not change but the column remains a relational reference.

The power of spreadsheets makes repetitive calculations very easy. In many operations, periods of time are normally new columns. Each element becomes a row item with changes in time becoming the columns.

All spreadsheet programs allow for changes in the appearance of the spreadsheet. Headings, borders, or type fonts are usual customizing tools.

NUMERICAL METHODS
This portion of numerical methods include techniques of finding roots of polynomials by Routh-Hurwitz criterion and Newton methods, Euler's techniques of numerical integration and the trapezoidal methods, and techniques of numerical solutions of differential equations.

Root Extraction

Routh-Hurwitz Method (Without Actual Numerical Results):

Root extraction, even for simple roots (i.e., without imaginary parts), can become quite tedious. Before attempting to find roots, one should first ascertain whether they are really needed or whether just knowing the area of location of these roots will suffice. If all that is needed is knowing whether the roots are all in the left half plane of the variable (such as in the s-plane when using Laplace transforms -- such as frequently the case in determining system stability in control systems), then one may use the Routh-Hurwitz criterion. This method is fast and easy even for higher ordered equations. As an example, consider the following polynomial:

$$P(x) = \prod_{m=1}^{n} (x - a_m) = x^n + a_1 x^{n-1} + a_2 x^{n-2} + \ldots + a_{n-1} \qquad (11\text{-}1)$$

Here, finding the roots for n>3 can become quite tedious without a computer; however if one only needs to know if any of the roots have positive real parts, then use the Routh-Hurwitz method. Here, an array may be made listing the coefficients of every other term starting with the highest power, n, on a line, then a following line may be made listing the coefficients of the terms left out of the first row. Following rows are constructed using Routh-Hurwitz techniques, and after completion of the array, one merely checks to see if all the signs are the same (unless there is a zero coefficient -- then something else needs to be done) in the first column; if none, no roots will exist in the right half plane. A simple technique used in control systems (for details, see almost any text dealing with stability of control systems. A short example follows:

$F(s) = s^3 + 3s^2 + 2s + 10,$

$= (s+?)(s+?)(s+?)$

Array:

	1	2
s^3	1	2
s^2	3	10
s^1	-4/3	0
s^0	10	0

Where the s^1 term is formed as: $(3 \times 2 - 10 \times 1)/3 = -4/3$. For details refer to any text on control systems or numerical methods.

Here, there are two sign changes, one from 3 to -4/3, and one from -4/3 to 10; this means there will be two roots in the right half plane of the s-plane, which yields an unstable system; this technique represents a great savings in one's time without having to actually factor the equation.

Newton's Method

The use of Newton's method of solving a polynomial and the use of iterative methods can greatly simplify the problem. This method utilizes synthetic division and is based upon the remainder theorem. This synthetic division requires estimating a root at the start, and, of course, the best estimate is the actual root. The root is the correct one when the remainder is zero. (There are several ways of estimating this root, including a slight modification of the Routh-Hurwitz criterion.)

By taking a $P_n(x)$ polynomial (see equation 11-1) and dividing it by an estimated factor $(x-x_1)$, the result is a reduced polynomial of degree n-1, $Q_{n-1}(x)$, plus a constant remainder of b_{n+1}. Thus, another way of describing equation 11-1 is,

$$P_n(x)/(x-x_1) = Q_{n-1}(x) + b_{n-1} / (x-x_1) \quad \text{or as} \quad P_n(x) = (x-x_1)Q_{n-1}(x) + b_{n-1} \tag{11-2}$$

If one lets $x=x_1$, the equation 11-2 becomes,

$$P_n(x=x_1) = (0)Q_{n+1}(x) + b_{n+1} = b_{n+1} . \tag{11-3}$$

Equation 11-3 leads directly to the remainder theorem: "The remainder on division by $(x-x_1)$ is the value of the polynomial at $x= x_1$, $P_n(x_1)$." *

Newton's method (actually, the Newton-Raphson method) for finding the roots for an nth order polynomial is an iterative process involving obtaining an estimated value of a root (leading to a simple computer program). The key to the process is getting the first estimate of a possible root; without getting too involved, recall the the coefficient of x^{n-1} represents the sum of all of the roots, and the last term represents the product of all n roots, then the first estimate can be "guessed" within a reasonable magnitude. After a first root is chosen, find the rate of change of the polynomial at the chosen value of the root to get the next closer value of the root, x_{n+1}. Thus the new root estimate is based on the last value chosen,

$$x_{n+1} = x_n - P(x_n)/P'(x_n), \quad \text{where } P'(x_n) = dP(x)/dx \text{ evaluated at } x=x_n \tag{11-4}$$

NUMERICAL INTEGRATION

Numerical integration routines are extremely useful in almost all simulation type programs, design of digital filters, theory of z-transforms, and almost any problem solution involving differential equations. And since digital computers have essentially replaced analog computers (which were almost true integration devices), the techniques of approximating integration are well developed. Several of the techniques are briefly reviewed below.

Euler's Method

For a simple first order differential equation, say $dx/dt + ax = af$, one could write the solution as a continuous integral or as an interval type one:

$$x(t) = \int^t [-ax(\tau)+af(\tau)]d\tau \tag{11-5a}$$

$$x(kT) = \int^{kT-T} [-ax+af]d\tau + \int_{kT-T}^{kT}[-ax+af]d\tau = x(kT-T)+A_{rect} \tag{11-5b}$$

Here, A_{rect} is the area of $(-ax+af)$ over the interval $(kT-T) < \tau < kT$. One now has a choice looking back over the rectangular area or or looking forward. The rectangular width is, of course, T. For the forward looking case presented in a first approximation for x_1 is**,

$$x_1(kT) = x_1(kT-T) + T[ax1(kT-T)+af(kT-T)T = (1-aT)x_1(kT-T) +aTf(kT-T) \tag{11-5c}$$

*Gerald & Wheatley, *Applied Numerical Analysis*, Addison-Wesley, 3rd ed., 1985.
**This method is as presented in Franklin & Powell, *Digital Control of Dynamic Systems*, Addison-Wesley, 1980, page 55.

Or, in general, for Euler's forward rectangle method, the integral may be approximated in its simplest form (using the notation $t_{K+1} - t_K$ for the width, instead of T which is $kT-T$) as,

$$\int_{t_K}^{t_{K+1}} x(\tau)d\tau \approx (t_{K+1} - t_K)x(t_K) \tag{11-6}$$

Trapezoidal Rule

This trapezoidal rule is based upon a straight line approximation between a function, f(t), at t_o and t_1. To find the area under the function, say a curve, is to evaluate the integral of the function between point a and b. The interval between these points are subdivided into subintervals; the area of each subinterval is approximated by a trapezoid between the end points. It will only be necessary to sum these individual trapezoids to get the whole area; by making the intervals all the same size, the solution will be simpler. For each interval of delta t (i.e., $t_{K+1} - t_K$), the area is then given by,

$$\boxed{\int_{t_K}^{t_{K+1}} x(\tau)d\tau \approx (1/2)(t_{K+1} - t_K)[x(t_{K+1}) + x(t_K)]} \tag{11-7}$$

This equation gives good results if the delta t's are small but it is for only one interval and is called the "local error". This error may be shown to be $-(1/12)(\text{delta } t)^3 f'(t=\xi_1)$, where ξ_1 is between t_o and t_1. For a larger "global error" it may be shown that,

$$\text{Global error} = -(1/12)(\text{delta } t)^3 [f'(\xi_1) + f'(\xi_2) + \ldots + f'(\xi_n)] \tag{11-8}$$

Following through on equation 11-8 allows one to predict the error for the trapezoidal integration; This technique is beyond the scope of this review or probably the examination; however for those interested, please refer to pages 249-250 of the previous mentioned reference to Gerald & Wheatley.

NUMERICAL SOLUTIONS OF DIFFERENTIAL EQUATIONS

This solution will be based upon a first order ordinary differential equations. However the method may be extended to higher ordered equations by converting them to a matrix of first ordered ones.

Integration routines produce values of system variables at specific points in time and update this information at each interval of delta time as T (delta $t = T = t_{k+1} - t$). Instead of a continuous function of time, $x(t)$, the variable x will be represented with discrete values such that $x(t)$ is represented by x_0, x_1, x_2, ..., x_n. Consider a simple differential equation as before, as (based upon Euler's method),

$dx/dt + ax = f(t)$.

Now assume the deta time periods, T, are fixed (not all routines use fixed step sizes), then one writes the continous equation as a difference equation where: $dx/dt \approx (x_{k+1} - x_k)/T = -ax_k + f_k$ or, solving for the updated value, x_{k+1},

$$(x_{k+1}) = x_k - Tax_k + Tf_k \quad \text{for fixed increments.} \tag{11-9a}$$

By knowing the first value of $x_{k=o}$ (or the initial condition), the solution may be achieved for as many "next values" of x_{k+1} as desired for some value of T. The difference equation may be programmed in almost any high level language on a digital computer; however, T must be small as compared to the shortest time constant of the equation (here, $1/a$).

The following equation (with the "f" term meaning a "function of" rather than as a "forcing function" term -- as used in equation 11-5a) is in a more general form of equation 11-9a. This equation is obtained by letting the notation (x_{k+1}) become $y[k+1)\Delta t]$ and is written (perhaps somewhat more confusing) as,

$$\boxed{y[(k+1)\Delta t] \quad y(k\Delta t) + \Delta t f[(y(k\Delta t), k \Delta t]]} \tag{11-9b}$$

Reduction of Differential Equation Order

To reduce the order of a linear time dependent differential equation, the following technique is used. For example, assume a second order one:

$$x'' + ax' + bx = f(t), \text{ define } x = x_1 \text{ and } x' = x_1' = x_2, \text{ then, } x_2' + ax_2 + bx_1 = f(t),$$

$$x_1' = x_2 \quad \text{<--- By definition.}$$
$$x_2' = -bx_1 + ax_2 + f(t)$$

This equation can be extended to higher order systems and, of course, be put in a matrix form (called the state variable form). And it can easily be set up as a matrix of first order difference equation for solving digitally.

Packaged Programs

Most currently available packaged simulation programs use algorithms not necessarily based upon Euler's methods but more advanced methods such as the Runge- Kutta method. Automatic variable step size methods like Milne's may also be used. However, as mentioned before, these routines are all built into the packaged programs and may be transparent to the user. The user of a specialized program may be without knowledge of the high level language being employed (except for certain modifications).

Glossary of Computer Terms

Accumulators	Registers which hold data, address, or instructions for further manipulations in the ALU
Address bus	Two way parallel path connecting processors and memory containing addresses
AI	Artificial Intelligence
Algorithm	A sequence of steps applied to a given data set which solve the intended problem
Alphanumeric data	Data containing the characters a,b,c...z,0,1,2,..9

ALU	Arithmetic and logic unit
ASCII	American Standard Code for Information Interchange ,7 bit/character (Pronounced AS-key)
Asynchronous	Form of communications which message data transfer is not synchronous with the basic transfer rate requiring start/stop protocol
Baud rate	Bits per second
BIOS	Basic input/output system
Bit	0 or 1
Buffer	Temporary storage device
Byte	8 bits
Cache Memory	Fast look-ahead memory connecting processors with memory offering faster access to often used data
Channel	Logic path for signals or data
CISC	Complex instruction-set control
Clock rate	cycles per second
Control bus	Separate physical path for control and status information
Control unit	Fetches, decodes instructions to control the operations of registers and ALU's
CPU	Central Processing Unit, primary processor
Data buffer	Temporary storage of data
Data bus	Separate physical path dedicated for data
Digital	Discrete level or valued quantification vs analog or continuous valued
Duplex Communication	Communications mode where data is transmitted in both directions, but only in one direction at any one time
Dynamic memory	Storage which must be continually hardware refreshed to remain
EBCDIC	Extended Binary Coded Decimal Interchange Code - 8 bits/character (Pronounced EB-see-dick)
EPROM	Erasable programmable read-only memory
Expert systems	Programs with AI which learn rules from external stimuli
Floppy disk	Removable disk media in various sizes, 5-¼", 3-½"
Flowchart	Graphical depiction of logic using shapes and lines
G-byte	Giga bytes 1,073,741,824 bytes or 2^{30}
Half-duplex communication	2-way communications path in which only 1 direction operates at a time (transmit or receive)
Handshaking	Communications protocol to start/stop data transfer
Hard disk	Disk which has non-removable media
Hardware	Physical elements of a system
Hexadecimal	Numbering system (base 16) uses 0-9,A,B,...F
Hierarchical database	Database organization containing hierarchy of indexes/keys to records
I/O	Input/Output devices such as terminal, keyboard, mouse, printer

IR	Instruction register
K-bytes	Kilo-bytes 1024 bytes or 2^{10}
LAN	Local Area Network
LIFO	Last In-First Out
LSI	Large scale integration
Main memory	That memory seen by CPU
M-bytes	Mega-bytes 1,048,5 76 bytes or 2^{20}
Memory	Generic term for random access storage
Microprocessor	Computer architecture with Central Processing Unit in one LSI chip
MODEM	Modulator-demodulator
MOS	Metal Oxide Semiconductor
Multiplexer	Device which switches several input sources one at a time to an output
Narrow band	Frequency domain reference to channel band width versus baseband
Nibbles	4 bits
Non-volatile memory	As opposed to volatile memory, does not need power to retain present state
Number systems	Method of representing computer data into human readable information
OCR	Optical Character Recognition
OS	Operating system
OS memory	Memory dedicated to the OS, not useable for other functions
Parallel interface	A character(8 bit) or word(16 bit) interface with as many wires as bits in interface plus data clock wire.
Parity	Method for detecting errors in data, 1 extra bit carried with data, even or odd, to make the sum of one bits in a data stream
PC	Program counter, or personal computer
Peripheral devices	Input/Output devices not contained in Main processing hardware
Program	A sequence of computer instructions
PROM	Programmable Read-Only Memory
Protocols	Established set of handshaking rules enabling communications
Pseudocode	An English-like way of representing structured programming control structures
RAM	Random Access Memory
Real time/Batch	Method of program execution; Real-time implies immediate execution, Batch mode is postponed until run on a group of related activities
Relational database	Database organization which related individual elements to each other without fixed hierarchical relationships
RISC	Reduced instruction set computer
ROM	Read Only Memory
Scratchpad memory	Highspeed memory either in hardware or software
Sequential Storage	Memory (usually tape) accessed only in sequential order (n, n + 1,..)

Serial Interface	Single data stream which encodes data by individual bit per clock period
Simplex communication	One-way communications
Software	Programmable logic
Stacks	Hardware memory organization implementing Last In-Last Out access
Static memory	Memory which does not require intermediate refresh cycles to retain state
Structured Programming	Use of Structured programming constructs such as Do-While, If-Then, Else
Synchronous	Communications mode which data and clock are at same rate
Transmission speed	Rate at which data is moved in baud. (bits per second,bps)
Virtual memory	Addressible memory outside physical address bus limits through use of memory mapped pages
Volatile memory	Memory whose contints are lost when power is removed.
VRAM	Video memory
Wide-band	300-3300 Hz
Words	8,16,or 32 bits
WYSIWYG	What you see is what you get
16-bit	Basic organization of data with 2-bytes per word
32-bit	Basic organization of data with 4-bytes per word
64-bit	Basic organization of data with 8-bytes per word
80386	Intel's microprocessor architecture based on 16 data address bus, extended vitual memory, external math coprocessor
80486	Upgrade to 80386 incorporating Math coprocessor within VLSI
80586	Intels microprocessor architector based on 16 bit data, 32 bit address bus

Problems and Solutions

11-1. In spreadsheets, what is the easier way to write
B1 + B2 + B3 + B4 + B5
(a) Sum(B1:B5)
(b) (B1..B5)Sum
(c) @B1..B5SUM
(d) @SUMB2..B5

11-2. The address of the cell located at column 23 and row C is
(a) 23C
(b) C23
(c) C.23
(d) 23.C

11-3 Which of the following is not correct?

(a) A CD-ROM may not be written to by a PC.
(b) Data stored in a batch processing mode is always up-to-date.
(c) The time needed to access data on a disk drive is the sum of the seek time, the head
        switch time, the rotational delay, and the data transfer time.
(d) The methods for storing files of data in secondary storage are: sequential file
organization, direct file organization, and indexed file organization.

11-4. Which of the following is false?
(a) Flowcharts use symbols to represent input/output, decision branches, process
statements and other operations.
(b) Pseudocode is an English-like description of a program.
(c) Pseudocode used symbols to represent steps in a program.
(d) Structured programming breaks a program into logical steps or calls to subprograms.

11-5. In Pseudocode using DOWHILE , the following is true:
(a) DOWHILE is normally used for decision branching.
(b) The DOWHILE test condition must be false to continue the loop.
(c) The DOWHILE test condition tests at the beginning of the loop.
(d) The DOWHILE test condition tests at the end of the loop.

11-6. A Spreadsheet contains the following formulas in the cells:

	A	B	C
1		A1+1	B1+1
2	A1^2	B1^2	C1^2
3	Sum(A1:A2)	Sum(B1:B2)	Sum(C1:C2)

If 2 is placed in cell A1, what is the value in Cell C3?
   (a) 12
   (b) 20
   (c)   8
   (d) 28

11-7. A matrix contains the following:

	A	B	C	D
1		3	4	5
2	2	A$2		
3	4			
4	6			

If you copy the formulas from B2 into D4, what is the equivalent formula in D4?
   (a) A$2
   (b) C4
   (c) C4
   (d) C$2

11-8. A Processing system is processor limiter when performing sorting of small tables in memory. Which would speed up computations?
   (a) Adding more main memory
   (b) Adding virtual memory
   (c) Adding cache memory
   (d) Adding peripheral memory

11-9. A small PC processing system is performing large (1MB) matrix operations which is currently I/O limited since memory is limited to 1Mbyte. Which would speed up computations?
   (a) Adding more main memory
   (b) Adding virtual memory
   (c) Adding cache memory
   (d) Adding peripheral memory

11-10. Transmission Protocol: Serial, asynchronous, 8 bit ASCII, 1 Start, 1 Stop, 1 Parity bit, 9600 bps. How long will it take for a 1Kbyte file to be transmitted through the link?
   (a) 0.85 sec.
   (b) 0.96 sec.
   (c) 1.07 sec.
   (d) 1.17 sec.

11-11. Transmission Protocol: Serial, synchronous, 8 bit ASCII, 1 Parity bit, 9600 bps. How long will it take for a 1Kbyte file to be transmitted through the link?

    (a) 0.85 sec.
    (b) 0.96 sec.
    (c) 1.07 sec.
    (d) 1.17 sec.

11-12. The binary representation 10101 corresponds to which base 10 number?

    (A) 3
    (B) 16
    (C) 21
    (D) 10,101

11-13. For the base 10 number of 30, the equivalent binary number is

    (A) 1110
    (B) 0111
    (C) 1111
    (D) 11110

11-14. On personal computers the data storage device with the largest storage capacity is most likely to be

    (A) 3½" diskette
    (B) hard disk
    (C) random access memory
    (D) 3½" HD diskette

11-15.

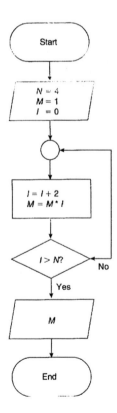

Figure 11-4.

The output value of M in Figure 11-4 is closest to
- (A) 1
- (B) 2
- (C) 8
- (D) 48

11-16. Pseudocode can best be described as
- (A) A simple letter-substitution method of encryption.
- (B) An English-like language representation of computer programming.
- (C) A relational operator in a database.
- (D) The way data are stored on a diskette.

11-17.

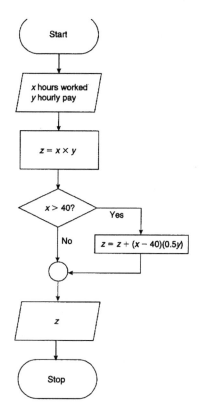

Figure 11-5

The pseudocode that best represents Figure 11-5 is:

(A)   INPUT X,Y
      WAGE = X x Y
      IF X less than 40 THEN
      WAGE = WAGE + OVERTIME PAY
      END IF
      OUTPUT WAGE

(B)   INPUT hours worked and hourly pay
      WAGE = hours worked x hourly pay
      IF hours worked greater than 40 THEN
      WAGE = WAGE + (hours worked -40)(overtime wage supplement)
      END IF
      OUTPUT WAGE

(C)   INPUT hours worked and hourly pay
      WAGE = (hours worked -40)(overtime rate) + 40(hourly pay)
      OUTPUT WAGE

(D)   PRINT hours worked and hourly pay
      WAGE = hours worked x hourly pay
      IF hours worked equals 40 THEN
      WAGE = hours worked x hourly pay + (hours worked -40) x overtime pay
      END IF
      OUTPUT WAGE

11-18.
All of the following is true of spreadsheets except:
   (A)   A cell may contain label, value, formula or function
   (B)   A cell reference is made by a numbered column and lettered row reference
   (C)   A cell content that is displayed is the result of a formula entered in that cell
   (D)   Line graphs, bar graphs, stacked bar graphs and pie charts are typical graphs created from spreadsheets.

Solutions

SOLN 11- 1

Sum(B1:B5) or @Sum(B1..B5) Answer is (a).

SOLN 11-2.

Answer is (b).

SOLN 11-3.

Batch mode processing always have delays in updating the database. Answer is (b).

SOLN 11-4.

Pseudocode does not use symbols but uses english-like statements such as IF-THEN, and DOWHILE. Answer is (c).

SOLN 11-5.

IFTHEN is normally used for branching. The DOWHILE test condition must be true to continue branching and the test is done at the beginning of the loop. The DOUNTIL test is done at the end of the loop. Answer is (c).

SOLN 11-6.

Plugging 2 into spreadsheet produces the following matrix:

	A	B	C
1	2	3	4
2	4	9	16
3	6	12	20

The value of C# is 20. Answer is (b).

SOLN 11-7.

The formula contains mixed references. The "$" implies absolute row reference while the column is relative. The result of any copy would eliminate any answer except for absolute row 2 entry relative column reference "A" gets replaced by "C". The cell contains C$2. The answer is (d).

SOLN 11-8.

Processor limited sorting on small talbes suggest either speeding up processor cycles or providing faster memory. Since speeding up clock is not an option, then making memory faster is the answer. Adding more memory, or virtual memory does nothing for small tables. Only adding cache memory would allow the CPU to fetch recently used data without full memory cycles thereby speeding up the sorting process. Answer is (c).

SOLN 11-9.
Processing is I/O limited since all of matrix can not fit into memory. Since this is a large matrix, cache memory probably would not effect processing. The best solution is adding more main memory to fit this problem entirely in memory. Answer is (a).

SOLN 11-10.
> 1 Kbyte=2^{10} bytes=1024
> 1024 bytes + 3 overhead bits/byte = 1024(8 bits/byte +3 bits/byte overhead)
> $\qquad\qquad$ = 11,264 bits
> Minimum transmission time = 11264 bits/9600 = 1.17 sec. Answer is (d).

SOLN 11-11.
> 1 Kbyte=2^{10} bytes=1024
> 1024 bytes + 1 overhead bit/byte = 1024(8 bits/byte + 1 bit/byte overhead)
> $\qquad\qquad$ = 9216 bits
> Minimum transmission time = 9216 bits/9600 = 0.96 sec. Answer is (b).

SOLN-11-12
The binary (base 2) number system representation is

	2^4	2^3	2^2	2^1	2^0	
	16	8	4	2	1	
Binary number	1	0	1	0	1	
	—	—	—	—	—	
	16	0	4	0	1	= 21

Answer is (C).

SOLN-11-13.

2^5	2^4	2^3	2^2	2^1	2^0	
32	16	8	4	2	1	
1	1	1	1	0	= 30	

Answer is (D).

SOLN-11-14. Answer is (B).

SOLN-11-15. Answer is (D).

SOLN-11-16. Answer is (B).

SOLN 11-17. Answer is (B).

SOLN-11-18.
A cell reference is made by a lettered column and a numbered row, example C3.
Answer is (B).

Chapter 12
PROCESS SAFETY

Process engineers are not only responsible for the most economical design and operation, they are also responsible for the safe design and operation of equipment. This chapter deals with the fundamentals of process safety.

Threshold Limit Values(TLV)

The threshold limit values are the highest concentration of a toxicant in air that can be tolerated for a given length of time without any adverse effect. There are five types of TLVs.

(1)TLV-TWA

The time weighted average of a toxicant for a normal eight-hour work day or 40-hour work week to which a worker may be repeatedly exposed ,day after day, without adverse effect.

(2)TLV-STEL

The short term exposure limit is the maximum concentration to which a person may be exposed continuously up to 15 minutes without adverse effect.

(3)TLV-C

The ceiling limit is the concentration which should not be exceeded any time.

(4) PEL

The permissible exposure limit is the maximum concentration exposure allowed by OSHA 29CFR 1910.1000

(5) IDLH

The concentration of the toxicant that is Immediately Dangerous to Life or Health is the maximum concentration from which one could escape within 30 minutes without escape-impairing symptoms or any irreversible health effects.

TLVs are expressed either in parts per million by volume(ppm) or milligram per cubic meter(mg/m^3). The relation between ppm by volume and mg/m^3 is as follows.

$$1 \ ppm \ = \ \frac{PM}{0.08205T} \ mg/m^3$$

Where:
P = pressure in atm., T = temperature in K, M= molecular mass.

Calculation of Time Weighted Average for Different Exposures to Same Toxicant

If a worker is exposed to different concentrations of the same toxicant at different lengths of time, the time weighted average of concentration is calculated as follows.

$$TWA \ = \ \frac{C_1 t_1 \ + \ C_2 t_2 \ +...+ \ C_n t_n}{t_1 \ + \ t_2 \ + \ ...+ \ t_n}$$

Where C_i is the concentration to which the worker is exposed for length of time t_i..

Procedure of Determining Overexposure When a Worker is Exposed to Environment Containing More Than One Toxicant.

The effect of more than one toxicant with different TLV-TWAs may be determined by evaluating the following value:

$$Value = \frac{C_1}{TLV_1} + \frac{C_2}{TLV_2} + \dots$$

Where C_1 is the concentration in ppm of the toxicant whose threshold limit value is TLV_1 (ppm) and so on. If the value of the above summation exceeds 1, then exposure limit is exceeded.

Procedure of Determining TLVs in Vapor Space of a Mixture of Liquids of Known Vapor Pressure and Composition

$$y_i = \frac{P^0_i x_i}{\Sigma P^0_i x_i}$$

$$TLV_{mixture} = \frac{1}{\Sigma \dfrac{y_i}{TLV_i}}$$

$$TLV_i \in mixture = y_i(TLV)_{mixture}$$

where:
P^0_i = pure component vapor presssure of component i at the mixture temperature
TLV_i = threshold limit value(ppm) of pure component i (when present by itself)
$TLV_{mixture}$ = threshold limit value of the vapor mixture of toxicants
$TLV_i \in mixture$ = TLV of component i in the mixture
x_i = mole fraction of component i in the mixture

Estimation of Dilution Air to Maintain Toxicant Concentration Below TLV
The requirement of dilution air to maintain toxicant concentration below TLV is given by:

$$Dilution\ air(CFM) = \frac{(3.87x10^8)W}{M(TLV)K}$$

Where W = flow rate or generation rate of toxicant, lb/min,
       M = molecular mass of toxicant

TLV = threshold limit value of toxicant, ppm

K = non-ideal mixing factor

= 1 for perfect mixing

= 0.1 to 0.5 for most real situation.

Fire Triangle

A fire is caused by the presence of sufficient quantity of each of the following three sources:

(1) Fuel : such as gasoline, hydrogen, wood, etc.,

(2) Oxidizer: such as oxygen, chlorine, ammonium nitrite, etc.,

(3) Ignition source: such as heat, mechanical impact, mechanical or electrical sparks or electrostatic discharge, etc.

Dust Explosion Pentagon

In addition to three sources mentioned above, two additional factors contribute to dust explosion:

(1) Confinement: This can be safeguarded by explosion venting.

(2) Dispersion.

Terms Used in Fire & Explosion

(1)Autoignition Temperature(AIT)

The temperature above which a flammable substance ignites itself by drawing sufficient amount of oxidizer and ignition from surrounding is called autoignition temperature.

(2)Flash Point(FP)

The lowest temperature at which a liquid gives off sufficient vapor to form an ignitable mixture with air near the surface of the liquid is called the flash point. At flash point, the combustion is brief.

(3)Fire Point

The lowest temperature at which a liquid gives off sufficient vapor to form an ignitable mixture with air, and sustains combustion once ignited, is called the fire point.

(4)Upper Flammability Limit(UFL)

The highest concentration of a flammable fluid in air above which the air-fuel mixture will not burn is called the upper flammability limit. The fuel concentration, by definition, is expressed as follows.

$$UFL = \frac{moles\ fuel}{moles\ fuel\ +\ moles\ air}$$

(5)Lower Flammability Limit(LFL)

The lowest concentration of a flammable fluid in air below which the air-fuel mixture will not burn is called the lower flammable limit. The concentration of fuel is expressed as in UFL.

(6)Explosion

The bursting or rupture of an enclosure or a container due to the development of internal pressure from a deflagration is called explosion.

(7)Deflagration

Propagation of combustion zone at a velocity that is less than the speed of sound in the unrecated medium is called deflagration.

(8)Detonation

Propagation of combustion zone at a velocity that is greater than the speed of sound in the unreacted medium is called detonation.

(9) Combustible Dust

   Any finely divided material, 420 microns or less in diameter(i.e., passing through US # 40 sieve), that presents a fire or explosion hazard when dispersed and ignited in air or other gaseous oxidizer is called combustible dust.

(10)Minimum Oxygen Concentration(MOC)

   The minimum per cent of oxygen in air plus fuel required to propagate flame is called the MOC. This is also used to signify minimum <u>oxidant</u> concentration, because substances other than oxygen can cause ignition. By definition, MOC is expressed by:

$$MOC = \frac{moles\ oxygen}{moles\ oxygen\ +\ moles\ fuel}$$

It follows from the definition of MOC and LFL that:

$$MOC = LFL\left(\frac{stoichiometric\ moles\ of\ oxygen}{1\ mole\ of\ fuel}\right)$$

The stoichiometric number of moles of oxygen needed to completely burn 1 mole of fuel to carbon dioxide and water can be determined from the stoichiometry of combustion reaction(parameter z in the following pages).

(11)Minimum Ignition Energy(MIE)

   The minimum energy in milli Joules required to start combustion is called MIE.

Classification of Liquids for Thermal Hazards Analysis

Flammable Liquid

A liquid having a closed cup flash point below 100 °F(37.8 °C) and having a vapor pressure not exceeding 40 psia at 100 °F is called flammable liquid. It is also known as Class I liquid. Flammable liquids are subdivided into three classes.

	Class IA	Class IB	Class IC
Flash point °F:	<73	< 73	≥ 73 < 100
Boiling point °F:	< 100	≥ 100	< 100
Venting device	CV & FA	CV or FA	CV or FA
for above-ground tanks:	Note 1	Note 1,2,3	Note 1,2,3

Notes:
(1) CV = Conservation vent, FA = Flame arrestor
(2)Tanks of 3000 bbl capacity or less containing crude petroleum in crude producing areas, and outside above-ground atmospheric tanks under 23.8 bbl capacity shall be permitted to have open vent.
(3)CV or FA may be ommitted where conditions such as plugging, crystallization, polymerization freezing, etc., may cause obstruction resulting in the damage of the tank. Consideration should be given to heating the devices, liquid seal, or inerting.

Combustible Liquid
A liquid having a closed cup flash point at or above 100 °F is known as combustible liquid. Combustible liquids are subdivided into three classes.

	Class II	Class IIIA	Class IIIB
Flash point °F	>100	≥140	≥200
	<140	<200	

Above-ground tanks larger than 285 bbl capacity storing Class IIIB liquids and not within the diked area or the drainage path of Class I or Class II liquids do not require emergency relief venting for fire exposure.

Computation of flash point of a solution containing a single flammable or combustible liquid dissolved in inerts.

(1) Find or experimentally determine the flash point of the flammable liquid in its pure state.
(2) Find the vapor pressure of the flammable liquid at the flash point. Call it p_i.
(3) Compute the pure component vapor pressure P^0_i by:

$$P^0_i = p_i / x_i$$

where x_i is the concentration in mole fraction of the flammable liquid in solution.
(4) Determine the temperature at which the vapor pressure of the flammable liquid equals P^0_i. This temperature is the flash point of the solution.

Procedure of Calculating Flammability Limits of a Mixture of Combustible Gases
The flammability limits may be reliably predicted by the Le Chatlier formulas.

$$LFL_{mixture} = \frac{1}{\dfrac{y_1}{LFL_1} + \dfrac{y_2}{LFL_2} + \ldots + \dfrac{y_n}{LFL_n}}$$

$$UFL_{mixture} = \frac{1}{\dfrac{y_1}{UFL_1} + \dfrac{y_2}{UFL_2} + \ldots + \dfrac{y_n}{UFL_n}}$$

where y = mole fraction of components in the mixture on a combustible basis or inert-free basis. This is calculated by dividing the % volume of the combustible component in the mixture by the %volume of all the combustible components. Subscripts denote the compnents. If the LFL_i and UFL_i in the denominator are entered as %, then the answer is also obtained as %. If the % volume of the total combustibles in the mixture including the inerts falls in the flammability range of the mixture(i.e., range of $LFL_{mixture}$ to $UFL_{mixture}$), then the mixture is flammable.

The dependence of flammability limit on temperature is expressed by the correlation of Zabetakis,

Lambiris & Scott:

Flammability Limit at T °C = Flammability Limit at 25 °C[1 ± 0.75(T - 25)/ΔH_c]

Where the + sign before 0.75 is to be used for UFL, and - sign for LFL. ΔH_c is net heat of combustion in kcal/mol.
Zabetakis also devluped the correlation of flammability limit and pressure.

$$UFL_p = UFL + 20.6(\log P + 1)$$

Where P = pressure in megapascals absolute = MPa, guage + 0.101
UFL = upper flammability limit at 1 atm.
Lower flammability limit has little sensitivity to pressure.

Stoichiometry of Combustion Reaction and Estimation of Flammability Limits.

The stoichiometry of the combustion reaction of hydrocarbons containing C, H, and O may be represented by:

$$C_mH_xO_y + z\,O_2 = m\,CO_2 + (x/2)H_2O$$

By stoichiometry:

$$z = m + x/4 - y/2 = \text{moles of oxygen per mole of fuel}$$

For hydrocarbon vapors, Jones formulas may be used to estimate flammability limits.

$$LFL = 0.55\,C_{st}$$

$$UFL = 3.5\,C_{st}$$

Where LFL and UFL are in % by volume and C_{st} is the stoichiometric concentration of fuel in the combustion reaction , also in % by volume, and is defined by:

$$
\begin{aligned}
C_{st} &= \frac{(moles\ fuel)100}{moles\ air\ + moles\ fuel} \\[2mm]
&= \frac{100}{\dfrac{moles\ air}{moles\ fuel} + 1} \\[2mm]
&= \frac{100}{\dfrac{moles\ air}{moles\ oxygen} \cdot x \dfrac{moles\ oxygen}{mole\ fuel} + 1} \\[2mm]
&= \frac{100}{\dfrac{z}{0.21} + 1}
\end{aligned}
$$

By combining above equations, it can be shown that:

$$LFL = \frac{55}{4.76m + 1.19x - 2.38y + 1}$$

$$UFL = \frac{350}{4.76m + 1.19x - 2.38y + 1}$$

Inerting
It is the technique of rendering a combustible mixture nonignitable by the addition of an inert gas.

Purging
It is the technique of rendering a combustible mixture nonignitable by the addition of an inert or combustible gas.

Three common methods of inerting are:
(1) Sweep through purging,
(2) Pressure purging
(3) Vacuum purging.

Sweep Through Purging

The purging gas is added to a vessel at one end and withdrawn at another end continuously. Assuming perfect mixing, with a continuous volumetric purge rate of Q (volume/time) through a container of volume V containing a toxicant , oxygen, or a combustible substance with a concentration C which is to be reduced in time t, one can write differential equation as follows:

$$- V\frac{dC}{dt} = QC$$

Separating variables, and integrating, one gets:

$$t = \frac{V}{Q}\ln\left(\frac{C_0}{C}\right)$$

The above equation shows purging time for perfect mixing in the vessel. In real situation, a mixing factor K is introduced for imperfect mixing. K ranges from 0.1 to 0.5 in most real situation.

$$t = \frac{V}{KQ}\ln\left(\frac{C_0}{C}\right)$$

Where C_0 is the concentration in the beginning, and C = concentration at the end.

Defining N = Qt/V = number of changes of container volume, one can write:

$$N = \frac{1}{K} \ln\left(\frac{C_0}{C}\right)$$

(2) Pressure purging & (3) Vacuum purging

Pressure and vacuum purging are batch operations. In pressure purging, a vessel is pressurized with an inert, and then vented usually to atmospheric pressure. In vacuum purging, the contents of the vessel is withdrawn by a vacuum system until a predetermined vacuum is reached, followed by an injection of an inert to the vessel to bring the pressure to the initial condition. In both cases the operation may be repeated until the desired concentration is reached. Both methods may be evaluated by the same formulas.

$$n = \frac{\ln\left(\frac{y_0}{y_n}\right)}{\ln\left(\frac{P_h}{P_l}\right)} \qquad M_i = \frac{n(P_h - P_l)MV}{RT}$$

Where:
M = molecular mass of inert, lb/lbmol
M_i = mass of inert required for purging, lb
P_h = high pressure of the cycle, psia
P_l = low pressure of the cycle, psia
n = number of purging cycles required
R = gas constant, 10.73 (psia.ft^3/lbmol.°R)
T = temperature of operation, °R
V = volume of container, ft^3
y_0, y_n = concentration of oxygen or combustible substance in the beginning and after n purge cycles, in consistent units.

References:
1."**Chemical Engineering License Review**", D. K. Das and R. K. Prabhudesai, Engineering Press, Austin, TX
2."**Chemical Process Safety: Fundamentals with Applications**", Daniel A. Crowl / Joseph F. Louver,Prentice Hall, Englewood Cliffs, NJ
3."**NFPA 69**" 1992 Edition
4."**Industrial Explosion Prevention and Protection**", Frank T. Bodurtha, McGraw-Hill Book Co.

PROBLEMS

Problem 12.1 Estimate the flash point of a solution containing 50% by mole of ethanol in water. The flash point of pure ethanol is 13 °C. The vapor pressure-temperature relationship of ethanol is given by:

$$\ln P_{mmHg} = 18.912 - \frac{3804}{T(K) - 41.68}$$

Problem 12.2 Estimate the MOC, LFL, and UFL of methanol in air.

Solutions

Problem 12.1

The flash point is 13 °C = 286 K. The vapor pressure of ethanol at 286 K is:

$$P_{ethanol\ @\ 286\ K} = e^{18.912 - \frac{3804}{286 - 41.68}} = 28.28\, mmHg$$

The pure component vapor pressure which corresponds to the partial pressure of 28.28 mmHg of a 50% solution of ethanol is given by:

$$P^0_{ethanol} = 28.28/0.5 = 56.56\ mmHg$$

The temperature which corresponds to the vapor pressure of 56.56 mmHg of ethanol is:

$$T = \frac{3804}{18.912 - \ln 56.56} + 41.68 = 297.38\ K = 24.38°C$$

Therefore, the flash point of the solution is 24.38°C.

Problem 12.2

Consider the stoichiometry of combustion reaction:

$$C_m H_x O_y + z\ O_2 = m\ CO_2 + (x/2)H_2O$$

Since methanol CH_3OH may be considered $C_1H_4O_1$,
m = 1, x = 4, y = 1, and hence z = m + x/4 - y/2 = 1.5. Therefore,

$$C_{st} = \frac{100}{4.76x1 + 1.19x4 - 2.38x1 + 1} = 12.92\ \%$$

$$LFL = 0.55C_{st} = 7.11\ \%$$

$$UFL = 3.5C_{st} = 3.5x12.92 = 45.22\ \%$$

$$MOC = LFL \left(\frac{stoichiometric\ moles\ of\ oxygen}{1\ mole\ of\ fuel} \right) = (LFL)z$$

$$= 7.11 \times 1.5 = 10.67\ \%$$

Chapter 13
Pollution Prevention(Waste Minimization)

Pollution is a by-product of civilization, because it is thermodynamically impossible to eliminate all wastes in a process involving transformation of matter and energy required to sustain a civilized society. In the name of civilization and progress, however, humans have a tendency to rationalize defecation in their own nests.

Philosophically, pollution is misplacement of a substance in a concentration or amount that is directly or indirectly hazardous to health or life in any form. Technically, only when such misplacement violates the local or federal laws, is it called pollution. In the latter connotation, misplacement of substances without violating laws does not necessarily mean a healthy atmosphere.

Pollution prevention should not be left as an "end-of-pipe" problem. It should be addressed at the stage of conceptual design. Careful consideration should be given to the waste treatment cost which is affected by the choice of solvents, starting raw materials, operating parameters and their mode of control, and work practices including detection, monitoring, reduction, and reporting. Most pollution can be prevented, or at least minimized with the best available control technology considered in the beginning of the conceptual design with support from management guided by enlightened self-interest.

Federal Pollution Prevention Act of 1990:
(1) Pollution should be prevented or reduced at the source wherever possible.
(2) Pollution that cannot be prevented should be recycled in an environmentally safe manner whenever feasible.
(3) Pollution that cannot be prevented or recycled should be treated in an environmentally safe manner whenever feasible.
(4) Disposal or other release into the environment should be considered only as a last resort and should be conducted in an environmentally safe manner.

Categories of Wastes:
(1) Municipal solid wastes such as paper, glass/metal, food, plastics etc. Disposal involves landfill, incineration , recycle etc.
(2) Industrial Hazardous wastes. Disposal involves recycle, landfill, incineration, land treatment, reuse as fuels, underground injection, waste piles, waste treatment and controlled discharge to natural water, and air.

Some commonly used terms in pollution prevention are explained below.

Acceptable Daily Intake(dose):
It is the number of milligrams of a chemical per kilogram of body weight which, taken daily during an entire life time, appears to be without risk on the basis of all known facts at the time.

Acid Rain:
All rainfalls are slightly acidic because of natural carbon dioxide. However, acid rain refers to the formation of acids due to the reaction of acidic oxides such as oxides of sulfur and oxides of nitrogen with rain water.

Activated Carbon:
Carbon obtained from vegetable and animal sources and roasted in a furnace to develop high surface area(1000 m^2/gm) is called activated carbon.

Activated Sludge:
The activated sludge is formed in the biological treatment of waste water which is mixed with biological culture(bacteria) in an aeration basin. Because of the growth of the bacteria and protozoa, part of the active sludge(about 90%) is incinerated, or disposed of as landfill and balance recycled.

Biochemical Oxygen Demand(BOD):
It is the number of milligrams of dissolved oxygen consumed by one liter of waste for a specified incubation period in days at 20°C. The incubation period is shown as a suffix. For example, BOD$_5$ denotes BOD with a 5-day incubation period.

Carcinogens:
Any substance that induces cancer to man or animal is called a carcinogen. A material is considered carcinogen if (1)it is certified as a carcinogen or potential carcinogen by IARC(International Agency for Research on Cancer),(2)it is listed as a carcinogen or potential carcinogen in the **Annual Report on Carcinogen** (3) it is regulated by OSHA as a carcinogen (4) a positive study has been published. Some industrially known carcinogens are asbestos, aldrin, benzene, beryllium, cadmium, carbon tetrachloride, chloroform, p-dichlorobenzene, dieldrin, DDT, formaldehyde, hexamethylenediamine, selenium, tetraethyl lead, toluene-24-diamine, trichloroethylene, vinyl chloride etc.

Chemical Oxygen Demand(COD):
It is the number of milligrams of oxygen which one liter of waste will absorb from a hot acidic solution of potassium dichromate.
The higher the values of BOD and COD, the more dangerous is the waste to aquatic life, because the waste depletes oxygen from the water required for sustaining life. A substance with low BOD may mean that substance is not biodegradable.

Fugitive Emission:
Fugitive emissions are unintentional releases from equipment such as pumps, valves, flanges, open-ended lines, filling losses, evaporation from sumps or ponds, etc. Environmental Protection Agency has five methods to estimate fugitive emission.

Greenhouse Effect:
Combustion of fossil fuel(coal, oil, natural gas) generates carbon dioxide(about 20 billion tons/yr) which forms a layer in the earth's atmosphere. This layer traps heat radiating from earth's surface thereby leading to global warming effect. Potential effect may be melting of polar ice leading to flooding in the coastal areas. Other contributors to the Greenhouse Effect

are CFC's , methane , and ozone. The role of chemical engineers here is to develop alternative fuels, refrigerants, and propellants..

Hazardous Air Pollutant(HAP):
Pollutants are hazardous if they are carcinogens or if exposure to them may cause serious health problem. The Clean Air Act Amendment of 1990 includes 189 HAPs[4].

Lethal Concentration & Dose:
A concentration or dose that causes death is called lethal concentration or dose. Lethal concentration, LC_{50} is the concentration of the substance in mg per liter that causes death within 96 hrs to 50% of the test group of the most sensitive important species in the locality under consideration. Lethal dose, LD_{50} is the dose of the substance in mg per kg of body weight of a specific animal (such as a rat) that causes death to 50% of the sample of the animals under test. A substance with LD_{50} greater than 7 g/kg is not considered harmful.[3]
.

Life Cycle Analysis:
Life cycle analysis of a product considers all impacts on environment starting from the procurement of the raw material, generation, use, and disposal of the product.

Mutagen:
Anything e.g. a chemical or radiation that changes the chromosomes of a cell so that the chromosomes of the daughter cells will be changed after cell division, is called a mutagen. A mutagen may be carcinogen or teratogen(an agent that causes birth defects, dioxin for example).

pH:
It is the negative logarithm to the base 10 of hydronium ion activity.

$$pH = -\log_{10}\left(a_{H_3O^+}\right) = -\log_{10}\left(\gamma C_{H_3O^+}\right) \cong -\log_{10}C_{H_3O^+} \cong -\log_{10}C_{H^+}$$

Where a = activity, γ = activity coefficient, usually 1 for dilute solution, C = concentration in mols per liter, H_3O^+ = hydronium ion , H^+ = hydrogen ion. pH for water based solutions is normally expressed in a scale of 0(1 normal acid) to 14(1 normal base). This scale is arbitrary, and therefore, pH could be negative(stronger than 1 normal acid) and also greater than 14(stronger than 1 normal base). pH of 7 indicates neutral solution. Because of base 10 logarithm scale, for example, a solution of pH at 6 is 10 times acidic compared with a solution of pH at 7. pH of pure water is 7 only at 25°C, and it decreases as the temperature increases because of increased dissociation. pH of a solution is affected by the nature of the solvent, and even by neutral salts. This is because the activity coefficient is dependent on the ionic strengths of all ions, not just hydronium ions. Thus:

$$\log_{10}\gamma_{H_3O^+} = \frac{-0.5I^{0.5}}{1 + 3I^{0.5}}$$

where I = ionic strength = $\frac{1}{2}\Sigma(C_i Z_i^2)$, where the subscript denotes the ionic species, and Z denotes the ionic charge.

Acceptable pH of various waters[1] are: irrigation(4.5 - 9), municipal secondary treatment waste, fresh water aquatic life and wildlife (6 - 9), public supply and recreational water(5 - 9), marine water(6.5 - 8.5).

Total Organic Carbon(TOC):

It is number of milligrams of soluble organic carbon per liter of waste. Before the analysis, the inorganic carbon materials are removed by acidification and sparging, and insoluble solids are filtered. The test involves analysis of carbon dioxide generated by burning the sample.

Volatile Organic Compounds(VOC):
Any compound of carbon, **excluding** carbon monoxide, carbon dioxide, carbonic acids, metallic carbides or carbonates, and ammonium carbonate, which participates in atmospheric photochemical reactions is termed a VOC.

OZONE, FRIEND AND FOE.

Ozone as a friend and its depletion.
Created naturally at a distance of 15 to 20 miles above the surface of the earth, ozone protects life on earth by absorbing a major portion of cancer-causing ultra-violet radiation and, in so doing, releases heat in the stratosphere. CFCs(Chlorofluorocarbons used as refrigerant, fire extinguisher, and propellant) escape undecomposed to the stratosphere where, bombarded with ultra-violet radiation, release halogens which attack ozone, breaking it down to oxygen. Nitrogen oxides released by supersonic jets also destroy ozone. Thus depletion of friendly ozone takes place, allowing more of the ultraviolet radiation to reach the surface of the earth.
  Chemistry of ozone as a friend involves a cyclic process of photochemical reaction of formation and decomposition of ozone:
(a) Formation of ozone:
$$O_2(g) + h\nu \rightarrow 2\ O(g)$$
$$O(g) + O_2(g) + M(g) \rightarrow O_3(g) + M^*(g) + heat$$
(b) Decomposition of ozone:
$$O_3(g) + h\nu \rightarrow O_2(g) + O(g)$$
$$O(g) + O(g) + M(g) \rightarrow O_2(g) + M^*(g) + heat$$
  Chemistry of ozone depletion depends on the type of pollutant. With a CFC it is as follows:
(a) Photolysis or light-induced rupture of carbon-chlorine bond and generation of atomic chlorine:
$$CF_xCl_{4-x}(g) + h\nu \rightarrow CF_xCl_{3-x}(g) + Cl(g)$$
(b) Reaction of atomic chlorine with ozone leading to the depletion of the latter through regeneration of atomic chlorine:
$$Cl(g) + O_3(g) \rightarrow ClO(g) + O_2(g)$$
$$ClO(g) + O(g) \rightarrow Cl(g) + O_2(g)$$
In above equations, $h\nu$ represents photon energy(h = Planck's constant, ν = frequency of radiation), M = N_2, O_2, or H_2O, M^* = molecule M with excess energy.

Ozone as a foe and its creation.

Ozone is extremely reactive and toxic. It causes bronchitis and is extremely dangerous to asthma sufferers. Ozone is the key component of **smog**. Ozone is created by the oxides of nitrogen which is produced by automobiles, planes, chemical plants, burners etc. Oxides of nitrogen decompose by the bombardment of infiltrated ultraviolet radiation to nascent oxygen which reacts with molecular oxygen to produce ozone. Ozone is a suicidal byproduct of civilization.

$$N_2(g) + O_2(g) \leftrightharpoons 2NO(g) \quad \text{[Inside internal combustion engine or furnace]}$$
$$2NO(g) + O_2(g) \leftrightharpoons 2NO_2(g) \quad \text{[Inside IC engine, furnace and outside]}$$
$$NO_2(g) + h\nu \rightarrow NO(g) + O(g) \quad \text{[Outside air]}$$
$$O(g) + O_2(g) + M(g) \rightarrow O_3(g) + M^*(g)$$

ABATEMENT OF AIR POLLUTION

VOC Control Technologies:

(1) Thermal Oxidation(Incineration). By this technology, the VOC-laden air is burnt at 1300 to 1800 °F using supplementary fuel and air, if required, to produce water and carbon dioxide.

(2) Catalytic Oxidation(Incineration). Same as above except a catalyst is used to lower the temperature of reaction to 700 - 900 °F consequently reducing the fuel consumption.

(3) Adsorption. This technology uses medium like activated carbon, zeolite etc., to adsorb VOC by weak intermolecular forces. The adsorbed VOC may be recovered by steam stripping or vacuum desorption and condensation.

(4) Absorption. By this method, a liquid solvent such as water, high boiling hydrocarbons, caustic solutions(for acidic VOC), or acid may be used in a venturi, packed/tray/spray columns to capture VOC.

(5) Condensation. This technique uses low temperature coolants like brine, cryogenic fluid, chilled water in a surface condenser. Cryogenic fluids like liquid nitrogen and carbon dioxide can be directly injected into the VOC stream to effect condensation (direct contact condensation).

(6) Waste heat recovery boiler. In this process, the VOC-laden air is burnt in boilers to generate steam.

(7) Flare. A flare, which is generally used to control pollution during process upsets, may also be used to burn VOC-laden air.

(8) Bio-degradation. This process uses soil or compost beds containing cultured microorganisms to convert VOCs into harmless components. The VOC-laden air must be dust free and humid.

(9) Membrane Separation. Semipermeable membranes are used to separate VOC from air due to selective permeability of the gases.

(10) Ultraviolet Oxidation Technology. By this technology, the VOC is converted to carbon dioxide and water by oxygen-based oxidants like ozone, peroxides, and radicals like OH and O in presence of ultra-violet light.

Other technologies applicable to pollutants other than a VOC:
There are many air-pollutants which are not a VOC such as carbon monoxide, carbon dioxide, halogens, halogen acids, nitrogen oxides, ammonia, sulfur compounds to name a few. Many treatment technologies outlined for VOC's also apply for pollution control of non-VOC's.
 Control of oxides of sulphur: Methods used are lime/limestone scrubbing, alkaline-fly-ash scrubbing, sodium carbonate/bicarbonate/hydroxide scrubbing, magnesium oxide scrubbing, sodium sulfite regenerative process(Wellman-Lord Process), citrate process which uses aqueous solution sodium sulfite, bisulfite, sulfate, thiosulfate, and polythionate buffered by citric acid, dimethyl aniline/xylidine process which uses the compounds as solvent to absorb sulfur dioxide, and finally the Claus process which uses hydrogen sulfide to react with sulfur dioxide to produce elemental sulfur.

ABATEMENT OF WATER POLLUTION
Classification of water pollutants.
(1) Floating pollutants such as oil.
(2) Suspended pollutants such as organic or inorganic solids.
(3) Dissolved pollutants such as acids, organic and inorganic substances.

Treatment Processes.
(1) Pretreatment such as air floatation, pH adjustment, coagulation, precipitation, flocculation.
(2) Clarification such as sedimentation and removal of suspended solids.
(3) Filtration such as cartridge filter, drum, and plate-and-frame filter.
(4) High Gradient Magnetic Separation(HGMS) for separation of magnetic, weakly magnetic or non-magnetic particles.
(5) Aerated lagoons used for treating small(< 1 million gallons/day) waste water with BOD(150-300 mg/L)
(6) Biological treatment of organic waste water using microorganism which converts the organics into carbon dioxide, water, and methane gas. Most common types of biological processes are activated sludge(aerobic digestion, converts organic matter to carbon dioxide and water), trickling filter, fixed activated sludge treatment(a hybrid of activated sludge and trickling bed),anaerobic digestion(converts organic matter to carbon dioxide and methane), rotating biological contactor, and nitrification. The nitrification is a biological treatment involving ammonium waste which is converted to nitrates by two types of bacteria: nitrosomonas which oxidizes ammonium to nitrite, and nitrobacter which oxidizes nitrites to nitrate.
    Air in the aerobic digestion may be replaced by pure oxygen(UNOX process).
(7) Carbon adsorption
(8) Ion exchange
(9) Reverse osmosis, electrodialysis, ultrafiltration
(10) Incineration
(11) Stripping as applied to ground water
(12) Liquid-liquid extraction
(13) Chlorination
(14) Ozone treatment for organic and inorganic waste
(15) Freeze concentration, evaporation, & crystallization
(16) Oxyphotolysis- a combination of ozone treatment and ultraviolet light used for toxic

organic like malathion
(17) Carbon-catalyzed hydrogen peroxide treatment to remove cyanide

TREATMENT OF CONTAMINATED SOIL
(1) Bioremediation: treatment of contaminated soil with microorganism
(2) Incineration

References:
(1) **Industrial Waste Water Management Handbook,** Hardam Sing Azad, McGraw-Hill Book Company, N.Y.

(2) **CURRENT AND POTENTIAL FUTURE INDUSTRIAL PRACTICES FOR REDUCING AND CONTROLLING VOLATILE ORGANIC COMPOUNDS,** Nik Mukhopadhyay, E.C. Moretti, American Institute of Chemical Engineers, N.Y.

(3) **Dictionary of Environmental Science and Technology,** Andrew Porteous, John Wiley & Sons.

(4)**Handbook of Health Hazard Control in the Chemical Process Industry**, pp 106-112, S. Lipton, J. Lynch, John Wiley & Sons, Inc.

Chapter 13 Problems
Pollution Prevention

Prob 13-1. Calculate the biochemical oxygen demand in kg of two thousand m³ of waste with BOD₅ of 100.

Prob 13-2. Calculate the fugitive emission from a pump, inlet, and outlet flanges when the screening concentration (ppmv) is 800. Use the following correlations:

Pump fugitive emission (kg/h) = 1.333 (10⁻⁵·³⁴)(ppmv)⁰·⁸⁹⁸
Flange fugitive emission (kg/h) = 0.918 (10⁻⁴·⁷³³)(ppmv)⁰·⁸¹⁸

Solutions

Prob 13-1.

$$\text{Biochemical oxygen demand} = 100\frac{mg}{l} \times 1000\frac{l}{m^3} \times 2 \times 10^3 m^3$$

$$= 200 \text{ kg}$$

Prob 13-2.

Emission from the pump seals = 1.22 (10⁻⁵·³⁴)(800)⁰·⁸⁹⁸=0.00246 kg/h
Emission from the pump flanges = 2(0.918)(10⁻⁴·⁷³³)(800)⁰·⁸¹⁸ = 0.00805 kg/h
Total emission = 0.0105 kg/h

CHAPTER 14
DISTILLATION

Distillation is the process of separating components from their liquid mixtures with the use of thermal energy as the separating medium. It takes advantage of the fact that at equilibrium, the concentration of the more volatile components in the vapor phase is greater than that in the liquid phase. The basic data required in solving distillation problems are the equilibrium composition relationships between the liquid and vapor phases of the system undergoing distillation. In this review, emphasis is placed on the binary(two component) systems.

IDEAL SYSTEMS AND RAOULT's LAW:

Ideal liquid mixtures obey Raoult's law which relates the partial pressure of a component in the vapor phase to the liquid phase composition by the expression

$$p_i = x_i P_i$$

Combining Dalton's law referred to in chapter 2 earlier with Raoult's law, the following expression can be obtained to describe mixtures of ideal vapors and liquids in equilibrium

$$P_t = \sum_1^n p_i = \sum_1^n y_i P_t = \sum_1^n x_i P_i$$

For a single component, $\quad y_i P_t = x_i P_i$

where
- p_i = partial pressure of component i in the vapor phase
- x_i = mol fraction of component i in liquid phase
- P_i = vapor pressure of pure component i at the system temperature
- y_i = mol fraction of component i in vapor phase
- P_t = total pressure of the system

Relative Volatility:

The relative volatility a_{ij} of a component i with respect to component j in a mixture is defined

by the relation

$$a_{ij} = \frac{y_i / x_i}{y_j / x_j}$$

For an ideal mixture, it is equal to the ratio of the vapor pressures or $\quad a_{ij} = P_i / P_j$

For a binary (two component) system, $y_i = 1 - y_j$ and $x_i = 1 - x_j$. Therefore,

$$a_{ij} = \left(\frac{y_i}{1-y_i}\right)\left(\frac{1-x_i}{x_i}\right)$$

In general,

$$a = \left(\frac{y}{1-y}\right)\left(\frac{1-x}{x}\right)$$

from which

$$y = \frac{ax}{1+(a-1)x} \quad \text{and} \quad x = \frac{y}{a+y(1-a)}$$

In terms of vapor pressures of two components in a binary system, the above relations can be expressed as

$$y_1 = \frac{P_1 x_1}{P_1 x_1 + P_2(1-x_1)} = \frac{P_1 x_1}{P_t} \quad \text{and } x_1 = \frac{P_t - P_2}{P_1 - P_2}$$

where P_1 and P_2 are the vapor pressures of the components 1 and 2 respectively.

Boiling Point and Equilibrium Diagrams:

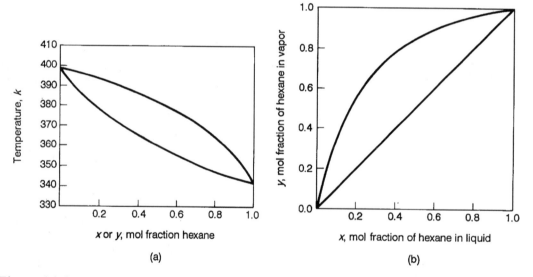

Figure 14-1.

For binary systems, the differences in composition of the liquid and vapor phases are graphically represented by either boiling point (temperature-composition) diagrams or only composition (equilibrium) diagrams. In the former, boiling point temperature is plotted against liquid and vapor composition whereas in the latter, the equilibrium vapor phase compositions of more volatile component are plotted against its corresponding liquid phase compositions as in Figure 14-la and b. For a liquid and vapor mixture at equilibrium, the ratio of the mols of the liquid phase to mols of the vapor phase is given by the inverse lever-rule as given below

$$\frac{L}{V} = \frac{y_{AV} - x_{AF}}{x_{AF} - x_{AL}}$$

where   L = moles of liquid phase
V = moles of vapor phase in equilibrium with the liquid
y_{AV} = mol fraction of component A in vapor phase.
x_{AF} = mol fraction of component A in the liquid feed before vaporization
x_{AL} = mol fraction of component A in the liquid phase at equilibrium.

For binary systems which obey Raoult's law and the vapor obeying Dalton's law of partial pressures, it is possible to calculate the boiling point and equilibrium diagrams from the vapor pressures of the pure components.

Non-ideal Binary Systems:

If the components show strong interactions (physical or chemical) then the boiling point and equilibrium diagrams are quite different from those of the ideal systems[1-6]. Azeotropic mixtures have a critical liquid composition at which the liquid and vapor compositions are the same. The partial pressures of components of a real solution will deviate from those calculated by Raoult's law. Non-ideal behavior may exist in either or both phases. For system pressures close to atmospheric, the real behavior may be accounted for by including a correction factor γ_i called the activity coefficient in Raoult's law. This results in the following relation

$$p_i = \gamma_i x_i P_i$$

The product $\gamma_i x_i$ is termed the activity a_i of the component i. The activity coefficient γ_i varies with temperature and composition and can be >1 or <1. By incorporating activity and fugacity coefficients v_i in Raoult's law, the following relation results

$$v_i y_i P_t = \gamma_i x_i P_i$$

With the use of this relation, the relative volatility in a non-ideal system is shown to be

$$a_{ij} = \frac{y_i / x_i}{y_j / x_j} = \frac{\gamma_i P_i v_j}{\gamma_j P_j v_i}$$

Many empirical and semi-empirical equations have been proposed to predict the variation of the activity coefficients with temperature and composition. Some of these are:

Van Laar Equations: These are as follows:

$$T \ln \gamma_1 = \frac{B}{(1 + A x_1 / x_2)^2} \quad \text{and} \quad T \ln \gamma_2 = \frac{AB}{(A + x_2 / x_1)^2}$$

where A and B are constants which are assumed to be independent of temperature. At the azeotropic composition,

$$x_i = y_i \text{ and } \gamma_i = P_t/P_i, \quad \text{Also } \gamma_i = y_iP_t/x_iP_i$$

Therefore knowing the azeotropic composition, the constants A and B can be calculated.

Margule's equations: These are given by:

$$\ln\gamma_1 = x_2^2[A + 2x_1(B-A)] \quad \text{and} \quad \ln\gamma_2 = x_1^2[B+2x_2(A-B)]$$

For a detailed treatment of activity coefficients and their estimation, the student is referred to other thermodynamic and distillation texts[1,3-8].

Equilibrium Vaporization Ratios (Equilibrium Constants) K:
K values are extremely useful in hydrocarbon distillation, absorption and stripping calculations. It is defined by the following relation:

$$K_i = \frac{y_i}{x_i}$$

In terms of vapor pressure, activity coefficient, and the fugacity coefficient, K_i is given by

$$K_i = \frac{\gamma_iP_i}{v_iP_t}$$

For an ideal system, both γ_i and v_i are unity and K_i reduces to $K_i = P_i/P_t$

Bubble Point:
The bubble point of a liquid mixture is the temperature at which it begins to boil at a given pressure such that the vapor is in equilibrium with the liquid. The condition can be expressed by the relation

$$\sum K_ix_i = \sum y_i = 1.0$$

for ideal systems,

$$\sum y_i = \sum \frac{P_ix_i}{P_t} = 1.0$$

Dew Point:
Dew point is the temperature at which the vapor when cooled begins to condense at a given pressure. This can be expressed in terms of equilibrium relationships as follows:

$$\sum \frac{y_i}{K_i} = \sum x_i = 1.0$$

For ideal systems,

$$\sum x_i = \sum \frac{y_i}{P_i}P_t$$

Flash (Equilibrium) Distillation:

This involves vaporization of a fraction of a batch of liquid mixture, keeping the liquid and the vapor in contact so that the vapor is in equilibrium with the liquid. The vapor is then withdrawn and condensed. This is a batch operation. In continuous flash distillation, the vapor and liquid are withdrawn from the flash chamber continuously. For a binary system, the process calculations can be done as follows

Over-all material balance: $\qquad F = V + L$

Component material balance: $\qquad Fx_F = Vy_i + Lx_i$

These two equations in conjunction with the equilibrium diagram allow us to determine the relative amounts of vapor and liquid produced. The energy balance for the process can be written as follows:

$$Fh_F = VH_v + Lh_L$$

where F = mols of feed, $\quad$ V = mols of flashed vapor, L = mols of undistilled liquid.
h_F, H_v, and h_L = molar enthalpies of feed, vapor and the condensed liquid.

For a multi component mixture, the following equations apply:

$$x_i = \frac{x_{Fi}/L}{K_i V/L + 1} \quad \text{or} \quad x_i = \frac{x_{Fi}/V}{K_i + L/V}$$

Since $y_i = K_i x_i$,

$$y_i = \frac{K_i x_{Fi}/V}{K_i + L/V}$$

and either $\sum x_i = 1.0$ or $\sum y_i = 1.0$. If one mol of the feed liquid is considered and f is

the fraction vaporized, then $(1-f)$ is the liquid left behind. Then by material balance,

$$x_{Fi} = f y_i + (1 - f) x_i$$

from which

$$y_i = -\frac{1-f}{f} x_i + \frac{x_{Fi}}{f}$$

For multi component (>2 components) mixtures, a trial and error solution is required. In case of the binary systems, the fraction vaporized is given by

$$f = \frac{x_{Fi}(K_1 - K_2)/(1 - K_2) - 1}{K_1 - 1}$$

Differential Batch Distillation:

In this type of distillation, the liquid mixture to be distilled is charged to a still and heated. The vapor is withdrawn and condensed continuously until a desired quantity of the distillate is obtained. The boiling point of the liquid in the still gradually increases as the lower boiling components vaporize. The Rayleigh equation relating final mols in the still L_2 to the initial charge L_1 mols of feed liquid is:

$$\ln\frac{L_1}{L_2} = \int_{x_{i2}}^{x_{i1}} \frac{dx_i}{y_i - x_i}$$

For a binary system, the above equation can be integrated graphically when x-y data are available. If relative volatilities α are constant, the direct solution of the equation is

$$\ln\frac{L_1}{L_2} = \frac{1}{\alpha - 1}\left(\ln\frac{x_{i1}}{x_{i2}} + \alpha\ln\frac{1 - x_{i2}}{1 - x_{i1}}\right)$$

For non-ideal systems,

$$y_i = \frac{\gamma_i x_i P_i}{v_i P_t} = K_i x_i$$

and

$$\ln\frac{L_1}{L_2} = \int_{x_{i2}}^{x_{i1}} \frac{dx_i}{x_i(K_i - 1)} = \int_{x_{i2}}^{x_{i1}} \frac{dx_i}{x_i(\gamma_i P_i / v_i P_t - 1)}$$

Fractional Distillation:

When higher product purities and large scale operation are required, continuous fractional distillation is employed. This involves contacting vapor and liquid in more than one stage, in a fractionating column (Figure 14-2a). Basic model used in the analysis of a fractionator is the equilibrium stage (Figure 14-2b). This means at steady state operation of the column, the liquid and vapor leaving the stage are in equilibrium. Number of theoretical stages required for a given separation are first computed. Actual number of stages is then determined by using a plate efficiency factor.

A portion of the condensed distillate called reflux is returned to the top plate in the column. This causes vaporization of the more volatile (light) component at each stage and condensation of the less volatile (heavy) component in the top or above the feed portion of the column. This enrichment of the more volatile component is called rectification and the portion of the column where it occurs is termed the rectification section.

In the section below the feed plate, hot vapor from the reboiler passes through each stage causing vaporization of the more volatile or light component and condensation of less volatile or heavy component. Thus the more volatile component is stripped from the liquid. This process is called stripping and the portion of the column where it is done is termed the stripping section of the column.

In multi component distillation, the light (more volatile) and the heavy (less volatile) components between which the required separation is specified are termed the key components (light key and heavy key). For the design of a column, the number of theoretical stages required for a given separation are first computed. Actual number of stages is then determined by using a plate efficiency factor. For the FE exam, only short, simple questions are asked, those generally referring to the McCabe-Thiele method given below.

Simple binary systems; McCabe -Thiele method:

Material and energy balances for a binary system are shown in Figure 14-2a. To simplify calculations, McCabe-Thiele analysis assumes constant molar overflows from stage to stage and the column operation at a constant pressure.

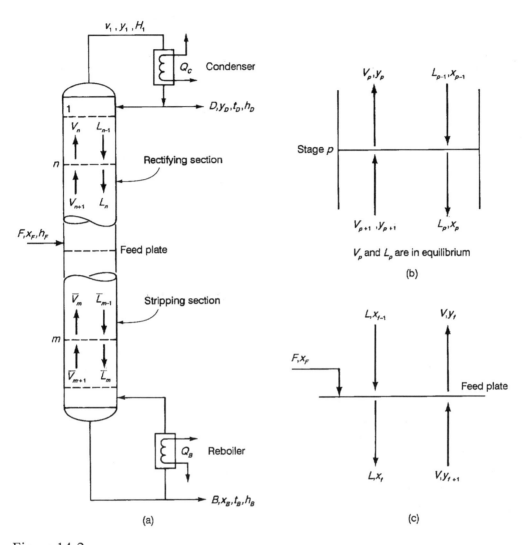

Figure 14-2.

Overall material balance: $\qquad$ $F = D + B$

Component A balance: $\qquad$ $Fx_F = Dx_D + Bx_B$

where F = feed to the column, mols/h $\qquad$ x_F = mol fraction of component A in feed

D = distillate, mol/h $\qquad$ x_D = mol fraction of component A in distillate

and B = bottoms product $\qquad$ x_B = mol fraction of component A in bottoms product

Rectifying section:

With reflux at its bubble point (total condenser), the operating line equation for the rectifying section of the column can be derived and is as follows

$$y_{n+1} = \frac{L}{V} x_n + \frac{D}{V} x_D$$

where L = liquid molar overflow from stage to stage, mol/h

   V = vapor molar flow from stage to stage, mol/h

   D = distillate product, mol/h

   x_D = mol fraction of more volatile component in the distillate

   x_n = mol fraction of component more volatile component in the liquid leaving stage n

and y_{n+1} = mol fraction of more volatile component in the vapor leaving stage n

The equation can also be written in terms of external reflux ratio R = L_o /D = L/D as:

$$y_{n+1} = \frac{R}{R+1} x_n + \frac{x_D}{R+1}$$

Slope of the operating line is

$$\frac{L}{V} = \frac{R}{R+1}$$

and the intercept of the line is x_D /(R+1).

Stripping section:

The corresponding equation of the operating line for stripping or bottom section of the column below the feed stage is

$$y_{m+1} = \frac{\overline{L}}{\overline{V}} x_n - \frac{B}{\overline{V}} x_B$$

where $\overline{L}$ = liquid overflow in the stripping section, mol / h

   $\overline{V}$ = vapor flow from stage to stage in the stripping section.

Since $\overline{V} = \overline{L} - \overline{B}$, the above equation becomes

$$y_{m+1} = \frac{\overline{L}}{\overline{L} - \overline{B}} x_n - \frac{B}{\overline{L} - \overline{B}}$$

The intersection of the operating lines occurs at a point which is determined by the feed composition and material   and energy balance at the feed plate. Using material balance at the feed plate (Figure 14-2c),

$$F + L + \overline{V} = V + \overline{L}$$

Now
$$\overline{L} = L + qF \quad \text{or} \quad V = \overline{V} + (1-q)F$$

where q is defined as: $q = \dfrac{\overline{L} - L}{F} = \dfrac{\text{heat to convert 1 mol of feed to saturated vapor}}{\text{latent heat of vaporization per mol of feed}}$

The equation of the stripping section operating line in terms of q is

$$y_{m+1} = \frac{L + qF}{L + qF - B} x_m - \frac{B}{L + qF - B} x_B$$

The equation of q-line can be obtained as:

$$y = \frac{q}{q-1} x - \frac{x_F}{q-1}$$

The q line has a slope of q/(q - 1) and terminates at x_F on the 45° line in the equilibrium diagram. It also locates the intersection point of the operating lines for all values of q.

q Values and q-line Position:
The position of the q line on the equilibrium diagram depends upon its value which is determined by thermal condition of the feed. The calculation of q is shown in Table 14-1. The effect of q value on the slope of q line is shown in Figure 14-3a and the method of the McCabe-Thiele graphical solution is shown in Figure 14-3b.

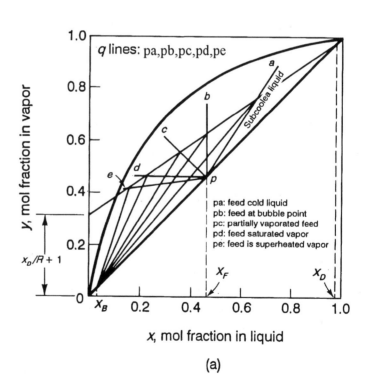

(a)

Figure 14-3a.

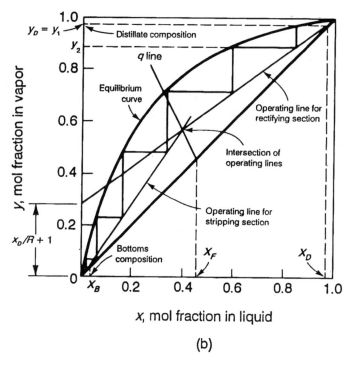

Figure 14-3b.

Table 14-1: calculation of q and slope of q line.

Condition of feed	Calculation of q	q	slope of q line q/(q-1)
Subcooled liquid	$q = 1 + C_{PL}(t_B - t_F)/\lambda$	$q > 1.0$	+
Feed at bubble point	$q = 1.0$	$q = 1.0$	∞
Partially vaporized feed	$q = f_L \lambda/\lambda = f_L$	$0 < q < 1.0$	-
Feed at dew point (Saturated vapor)	$q = 0$	$q = 0$	0
Feed superheated vapor	$q = \dfrac{C_{PV}(t_D - t_F)}{\lambda}$	$q < 0$	+

In the above table, the variables are as follows:

C_{PL} = molar specific heat of liquid
C_{PV} = molar specific heat of superheated vapor feed.
t_D = dew point temperature
t_B = boiling point
t_F = feed temperature
f_L = molar liquid fraction
λ = latent heat of vaporization per mole

Feed Plate Location:

The optimum location of the feed plate gives the least number of equilibrium stages required for a specified separation and reflux ratio. Least number of stages are obtained if the operating line switch from rectifying to stripping section is made at the intersection of the two operating lines.

Minimum Reflux:

Minimum reflux ratio corresponds to infinite number of equilibrium stages. It is a conceptual limit never actually attainable. As the reflux ratio is decreased, the intersection point of the operating lines gets closer to the equilibrium line until finally the intersection point of the operating line and the q-line lies on the equilibrium curve (Figure 14-4a) or one of the operating lines becomes tangent to the equilibrium curve (Figure 14-4b). In both cases, a pinch point is reached and an infinite number of equilibrium stages is required. These conditions correspond to minimum reflux which can be calculated from the slope of the critical operating line or from the y-axis intercept. as follows:

$$m = \frac{x_D - y_P}{x_D - x_P} = \left(\frac{L}{V}\right)_{min} = \frac{R_{min}}{R_{min} + 1} \quad \text{and} \quad R_{min} = \frac{m}{1 - m}$$

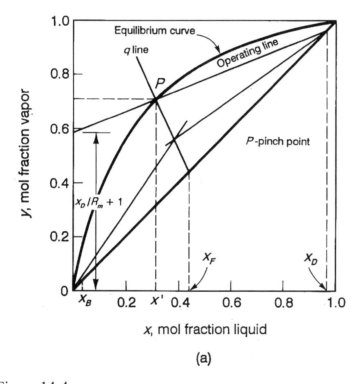

(a)

Figure 14-4a.

Total Reflux:

At total reflux, the operating lines for both sections of the column coincide with the 45° line. Then the number of equilibrium stages required is minimum, but there is no product. The minimum number of theoretical stages can be obtained graphically on the x-y diagram between compositions x_D and x_B using the 45° line as the operating line in both the rectifying and stripping sections. For ideal mixtures and constant relative volatility, the minimum number of theoretical plates N_{min} can be calculated by the Fenske equation.

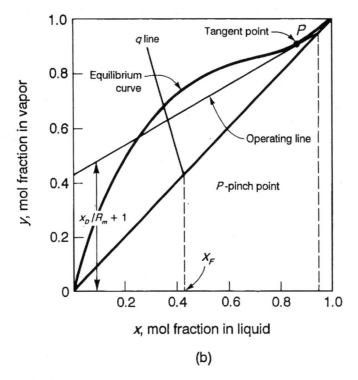

Figure 14-4b.

$$N_{min} = \frac{\log[x_D(1-x_B)/x_B(1-x_D)]}{\log\alpha_{AB}} - 1$$

If α changes moderately over the column, an average relative volatility can be used as follows:

$$\alpha_{av} = \sqrt{\alpha_{top}\alpha_{bottom}}$$

Partial Condenser:

A partial condenser is one in which a portion of the vapor is condensed. These are used when a vapor product is desired or if the temperature required to condense the vapor is too low. Equilibrium condensation occurs if the time of contact of the liquid and vapor is adequate. In this case, the condenser is equivalent to a theoretical equilibrium stage. In the design of new equipment, it is safer to ignore the enrichment by the partial condensation and include an additional theoretical tray in the column itself. **FE EXAM TIP: Usually the exam will assume a total condenser.**

Reboiler:

The reboiler is usually considered as one equilibrium stage. However, it depends upon the configuration of the reboiler. If the liquid is partially vaporized in the reboiler and fed to the bottom of the column, a vapor and liquid in equilibrium are produced. In this case, the reboiler can be considered equivalent to one equilibrium stage. When reboiler is used to boil a saturated liquid to a saturated vapor, no separation occurs and the reboiler is equivalent to a zero equilibrium stage. In case of a thermo-siphon reboiler, the effect is of a fraction of an equilibrium stage. For safety in design, the thermo-siphon reboiler is counted as zero stage.

Optimum Reflux Ratio:

As the reflux ratio is increased from the theoretical minimum, the total cost of running a distillation system consisting of the depreciation of the installed equipment, and the operating cost of coolant, heating medium, and electricity first decreases, reaches a minimum value and then increases. The reflux ratio at which the total cost is minimum is called the optimum reflux ratio. In actual practice, a reflux ratio equal to 1.2 to 2 times the minimum is used. The total cost is not very sensitive in this range and a better flexibility in operation is obtained.

Column Efficiency:

To obtain actual number of trays required in a column, a tray efficiency factor is required. Murphree plate efficiency is calculated by

$$E_0 = \frac{y_n - y_{n+1}}{y_n^* - y_{n+1}}$$

where E_0 = plate efficiency factor

y^*_n = vapor composition in equilibrium with L_n

y_n, y_{n+1} = actual vapor compositions from nth and (n + 1)th plates

Plate efficiencies are to be determined experimentally but some prediction methods are also available. O'Connell[9] gave an empirical correlation of plate efficiency as a function of feed viscosity and the relative volatilities of the key components. **FE Exam Tip:** Divide the theoretical number of plates by the efficiency to get the actual number of plates.

Column Energy Balance:

Neglecting the heat loss, the overall energy balance around the whole column gives

$$Fh_F + Q_B = Dh_D + Bh_B + Q_C$$

Condenser duty is given by $\quad Q_C = D(R + 1)(H_1 - h_D)$

where h and H are enthalpies per unit mol. Subscripts B, D, F, and 1 indicate bottom, distillate, feed, and tray l respectively. If condensate is at its bubble point (no subcooling),

$$Q_C = D(R + 1) \lambda$$

where λ latent heat of condensation of overhead vapor per mol.

If t_F is taken as the datum temperature, the feed is 100 % liquid, and t_D and t_B are the distillate and bottoms product temperatures respectively,

$$h_F = 0, \quad h_D = C_{PD}(t_D - t_F), \quad \text{and} \quad h_B = C_{PB}(t_B - t_F)$$

where C_{PD} and C_{PB} are the molar specific heats of the distillate and bottoms product.

Then reboiler duty is given by

$$Q_B = DC_{PD}(t_D - t_F) + BC_{PB}(t_B - t_F) + Q_C$$

Example Problems: FE EXAM TIP: the exam will usually include several questions on distillation using the McCabe-Thiele diagram.

Distill
14-1& 14-2:
At a temperature of 366.4 K, the vapor pressures of n-hexane and octane are 1480 and 278 mmHg respectively. Assume hexane-octane system obeys Raoult's law and the total pressure is 1 atm.

14-1: Calculate the equilibrium liquid composition of the more volatile component.

14-2: Calculate the equilibrium vapor composition of the more volatile component.

Distill
14-3-5:
A mixture of 50 mol% n-pentane and 50 mol% n-hexane is fractionated in a column to produce overhead product containing 95 mol % pentane and a bottom product containing 5 mol % pentane. The distillation is carried out at 1 atmosphere. The equilibrium diagram is shown in the figure 14-5.The feed enters the column at its bubble point.Total condenser is used.

14-3: Calculate the minimum reflux ratio

14-4: Determine minimum number of theoretical stages required, assume reboiler as a theoretical stage.

14-5: Calculate minimum number of theoretical plates using Fenske equation.

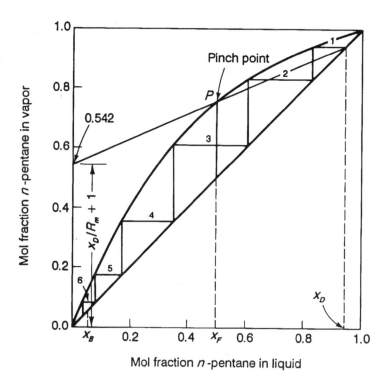

Figure 14-5.

Distill

14-6-9:

A fractionating column is to be designed to separate 5000 kg/h of mixture of 40 mol % benzene and 60 mol % toluene to produce a distillate containing 97 mol % benzene and a bottoms product containing 98 mol % toluene. The column is to be operated at one atmosphere and total condenser is to be used. The other data are as follows:

Bubble point of feed liquid = 368 K
Feed temperature = 293 K, feed specific heat = 1.6 x 10^2 kJ/(kmol.K)
Molal latent heat of vaporisation for benzene = 3.08 x 10^4 kJ/kmol
Molal latent heat of vaporisation for toluene = 3.33 x 10^4 kJ/kmol

Equilibrium diagram for benzene- Toluene system is given in Figure 14-6.
Calculate or obtain the following

14-6:   the slope of q line
14-7:   minimum reflux ratio
14-8:   number of theoretical stages required if reflux ratio 1.5 times the minimum is used.
14-9:   feed plate location

Distill 14-10-11:

In Problem 14-6 above,

14-10:   establish material balance, condenser duty and reboiler duty. Distillate Temperature = 355K, Bottom temperature = 382.8K. Neglect heat losses.
14-11:   calculate liquid and vapor molal overflows in the rectifying and stripping section.

Solutions to problems:

Distill 14-1:

By Raoult's law, $p_1 = x_1 P_1$ and $p_2 = (1 - x_1)P_2$

$$P_t = P_1 + P_2 = x_1 P_1 + (1 - x_1)P_2$$

Hence $$x_1 = \frac{P_t - P_2}{P_1 - P_2}$$

Using given values of vapor pressures, $x_1 = \frac{760 - 278}{1480 - 278} = 0.401$

14-2 :   and $y_1 = \frac{x_1 P_1}{P_t} = \frac{(0.401)(1480)}{760} = 0.781$

Note:

Similar calculations can be done at other temperatures with corresponding vapor pressures to construct T-x-y and x-y diagrams.

Distill 14-3 :

14-3: $x_F = 0.5$, $x_D = 0.95$, $x_B = 0.05$ L

Locate these points on the equilibrium diagram (Figure 14-5). Feed is at boiling point. Therefore q- line is vertical. It passes through the intersection of 45° line and the vertical from $x_F = 0.5$. Draw the q - line to intersect equilibrium line. This point is the pinch point. From (x_D, y_D), draw a line to pass through the pinch point to cut y axis to give an intercept = 0.542.

intercept $= \dfrac{x_D}{R_{m+1}}$ **= 0.542 Then $R_m + 1$ = 0.95 / 0.542 R_m = 1.75 - 1 = 0.75**

Minimum reflux ratio is 0.75

Distill 14-4:
Minimum number of stages are obtained when the operating lines coincide with 45° line on the equilibrium diagram. Step off the plates from $x_D = 0.95$. Number of stages is 6.6.
(Figure 14-5) Assume reboiler is one theoretical plate.
Then minimum number of plates = 6.6 - 1 = 5.6

Distill 14-5:

$$\alpha = \frac{y}{y-1}\left(\frac{1-x}{x}\right)$$

From figure 14- 5, $x_D = 0.95$, equilibrium composition of vapor = $y_0 = 0.98$
$x_B = 0.05$ equilibrium composition of vapor $= y_B = 0.11$

$$\alpha_{top} = \frac{0.98}{1-0.98}\left(\frac{1-0.95}{0.95}\right) = 2.58 \qquad \alpha_{bottom} = \frac{0.11}{1-0.11}\left(\frac{1-0.5}{0.05}\right) = 2.35$$

$$\alpha_{av} = \sqrt{2.58 \times 2.35} = 2.46$$

$$N_{min} = \frac{\log\left[0.95(1-0.05)/(0.05(1-0.95))\right]}{\log 2.46} - 1 = 5.5 \text{ theoretical plates by Fenske equation.}$$

Distill 14-6:

Heat required to raise 1 mol of feed to 368 K= 1 x $(1.6 \times 10^2)(368 - 293)$
$= 12000$ kJ/kg-mol
λ of feed mixture = $0.4(3.08 \times 10^4) + 0.6(3.33 \times 10^4) = 3.23 \times 10^4$ kJ/kg.mol
Therefore,

$$q = \frac{12000 + 3.23 \times 10^4}{3.23 \times 10^4} = 1.37$$

$$\text{slope of q - line} = \frac{q}{q-1} = \frac{1.37}{1.37-1} = 3.7$$

Distill

14-7: Refer to Figure 14-7 for solution.

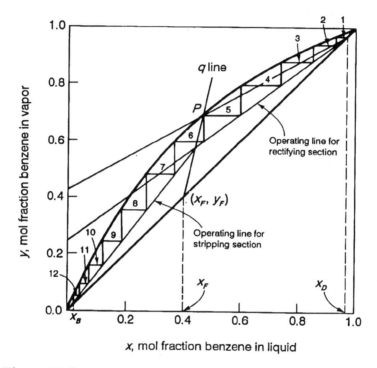

Figure 14-7.

First draw the q-line with slope of 3.7 and passing through the intersection point $(x_F y_F)$ of vertical from x_F and 45° line to intersect the equilibrium line at pinch point P. Join $(x_D\ y_D)$ and P and produce to meet y axis at intercept = 0.425.

Thus $\dfrac{x_D}{R_m + 1} = 0.425$ from which $R_m = 1.28$

Minimum reflux ratio is 1.28

Distill 14-8:

Actual reflux ratio = 1.5 x 1.28 = 1.92

Intercept of rectifying section operating line on y axis $= \dfrac{x_D}{R+1} = \dfrac{0.97}{1.92+1} = 0.25$

Operating line for rectifying section is drawn from (x_D y_D) to intersect y axis at y = 0.25 (Figure 14-7). Draw the operating line for stripping section. Step off the stages in usual manner.
Number of theoretical plates = 12.1 - 1 (Reboiler) = 11.1.
Distill
14-9: Crossover from one operating line to the other occurs at 6th plate.

Feed plate = 6.
Distill
14-10:

Material balance(overall): F = D+ B
Component balance: $Fx_F = Dx_D + Bx_B$
Molecular weight of feed = 0.4(78) + 0.6(92) = 86.4
Feed = 5000/86.4 = 57.87 kmols/h $x_D = 0.97$ $x_B = 0.02$

Substitute these values in the material balance equations and get the following
$$D + B = 57.87$$
$$D(0.97) + B(0.02) = 57.9(0.4)$$
Solving, D = 23.16   B = 34.71 kmols/h respectively.

Condenser duty:
Assume no subcooling.
Vapor flow from the column to the condenser = 1.92x23.16 + 23.16 = 67.63 kmol/h
λ = 0.97(3.08x10^4) + 0.03(3.33x10^4) = 3.09x10^4 kJ/kmol of distillate.
Then Q_c = 67.63x(3.09x 10^4) = 2.09x 10^6 kJ/h

Reboiler duty:
Assume molar specific heat of liquid is independent of composition.
Q_{SD} = Enthalpy of distillate relative to feed = 23.16(355 - 293)(1.6x10^2)
$$=2.3x10^5 \text{ kJ/h}$$
Q_{SB} = Enthalpy of bottoms relative to feed= 34.71(382.8 - 293)(1.6x 10^2)
$$= 4.99x 10^5 \text{ kJ/h}$$
Reboller duty =Q_c+ Q_{SD} + Q_{SB}= 2.09x10^6 + 2.3x10^5 + 4.99x10^5
$$= 2.82x10^6 \text{ kJ/h}$$
Distill
14-11: Vapor and liquid flows in column:

Flows from plate to plate in rectifying section:

Liquid flow: L = L_0 =RD= 1.92 D = 1.92(23.16) = 44.47 kmol/h
Vapor flow: V = L_0 + D = D(R + 1) = 23.16(1.92 + 1) = 67.63 kmol/h

Flows in stripping section:

$$\overline{L} = L + qF = 44.47 + 1.37(57.87) = 123.75 \text{ kmol/h}$$
$$\overline{V} = V - (1 - q)F = 67.63 - (1 - 1.37)(57.87) = 89.04 \text{ kmol/h}$$

Referencies:
1. Van Winkle M, **Distillation,** McGraw-Hill, New York, (I 967)
2. Das, D. K. and R. K.Prabhudesai, **Chemical Engineering License Review,** Engineering Press(1996)
3. Henley E.J.(Ed.), **Binary Distillation,** A. 1. Ch. E Modular Instruction, series B, vol 1(1980)
4. Perry R. H. (ed.), **Cheml Engrs' Hand Book,** 6th Ed. McGraw-Hill., New York, (1984)
5. Treybal R. E., **Mass Transfer Operations,** McGraw-Hill, New York, (1980)
6. Smith J. M., and H. C. Van Ness, **Introduction to Chem. Engr. thermodynamics, 3rd Ed.,** McGraw-Hill, New York, (I 975)
7. Hougen 0. A. et al, **Chemical Process Principles, Part 11,Thermodynamics,** John Wiley & Sons, New York (1959)
8. Robinson C. S. and E. R. Gilliland, **Elements of Fractional Distillation,** McGraw-Hill, New York(1950)
9. O'Connell H. C., **Trans. Am. Inst. Chem Engrs,** 42:751(1946)
10. Henley E.J., and J.M.Calo (ed), **Multicomponent distillation,** A.I. Ch.E Modular Instruction, Series B (1981)

AFTERNOON SAMPLE EXAMINATION

Chemical	No.Probs..
Material & Energy Balances	9
Chemical Thermodynamics	6
Mass Transfer	6
Chemical Reaction Engineering	6
Process Design and Econ Evaluation	6
Heat Transfer	6
Transport Phenomenon	6
Process Control	3
Process Equipment Design	3
Computer and Numerical Methods	3
Process Safety	3
Pollution Prevention	3

Instructions for Afternoon Session

1. You have four hours to work on the afternoon session. You may use the *Fundamentals of Engineering Reference Handbook* as your *only* reference. Do not write in this handbook.

2. Answer every question. There is no penalty for guessing.

3. Work rapidly and use your time effectively. If you do not know the correct answer, skip it and return to it later.

4. Some problems are presented in both metric and English units. Solve either problem.

5. Mark your answer sheet carefully. Fill in the answer space completely. No marks on the workbook will be evaluated. Multiple answers receive no credit. If you make a mistake, erase completely.

Work 60 afternoon problems in four hours.

FUNDAMENTALS OF ENGINEERING EXAM

AFTERNOON SESSION

(A)(B)(C) Fill in the circle that matches your exam booklet

1 (A)(B)(C)(D)	16 (A)(B)(C)(D)	31 (A)(B)(C)(D)	46 (A)(B)(C)(D)
2 (A)(B)(C)(D)	17 (A)(B)(C)(D)	32 (A)(B)(C)(D)	47 (A)(B)(C)(D)
3 (A)(B)(C)(D)	18 (A)(B)(C)(D)	33 (A)(B)(C)(D)	48 (A)(B)(C)(D)
4 (A)(B)(C)(D)	19 (A)(B)(C)(D)	34 (A)(B)(C)(D)	49 (A)(B)(C)(D)
5 (A)(B)(C)(D)	20 (A)(B)(C)(D)	35 (A)(B)(C)(D)	50 (A)(B)(C)(D)
6 (A)(B)(C)(D)	21 (A)(B)(C)(D)	36 (A)(B)(C)(D)	51 (A)(B)(C)(D)
7 (A)(B)(C)(D)	22 (A)(B)(C)(D)	37 (A)(B)(C)(D)	52 (A)(B)(C)(D)
8 (A)(B)(C)(D)	23 (A)(B)(C)(D)	38 (A)(B)(C)(D)	53 (A)(B)(C)(D)
9 (A)(B)(C)(D)	24 (A)(B)(C)(D)	39 (A)(B)(C)(D)	54 (A)(B)(C)(D)
10 (A)(B)(C)(D)	25 (A)(B)(C)(D)	40 (A)(B)(C)(D)	55 (A)(B)(C)(D)
11 (A)(B)(C)(D)	26 (A)(B)(C)(D)	41 (A)(B)(C)(D)	56 (A)(B)(C)(D)
12 (A)(B)(C)(D)	27 (A)(B)(C)(D)	42 (A)(B)(C)(D)	57 (A)(B)(C)(D)
13 (A)(B)(C)(D)	28 (A)(B)(C)(D)	43 (A)(B)(C)(D)	58 (A)(B)(C)(D)
14 (A)(B)(C)(D)	29 (A)(B)(C)(D)	44 (A)(B)(C)(D)	59 (A)(B)(C)(D)
15 (A)(B)(C)(D)	30 (A)(B)(C)(D)	45 (A)(B)(C)(D)	60 (A)(B)(C)(D)

DO NOT WRITE IN BLANK AREAS

Sample Exam

Prob 1-3: Iron and steam react according to the following equation

$$3Fe + 4 H_2O \rightarrow Fe_3O_4 + 4H_2$$

Assuming no excess reactants are used and the reaction is complete

Prob 1 The amount of iron (kg) required to produce 100 kg of hydrogen is
  a) 168 b) 2900 c) 2100 d) 1050

Prob 2: The amount of iron oxide (kg) produced is
  a) 2320 b) 2900 c) 2100 d) 1450

Prob 3: The volume (m³), H_2 will occupy at standard conditions if it were an ideal gas is
  a) 2240 b) 1120 c) 1184 d) 17950

Prob 4: Vapor pressures of Ethyl acetate are 3.22 kpa and 845.8 kPa at 0 and 160 °C
respectively. Its vapor pressure at 100 °C (kPa) is
  a) 183.2 b) 1.87 c) 137.9 d) 27.1

Prob 5-6: 10 kmols of the fuel gas (composition by volume:Methane 85%,Ethane 10.5% Nitrogen 4.5%)
is burned with 15 % excess air. If complete combustion is assumed,

Prob 5: Total number of moles after combustion per 10 kmols of gas is
  a) 1236.4 b) 272 c) 33.9 d) 124

Prob 6: If the total pressure is 100.6 kPa , the partial pressure of water vapor (kPa) is
  a) 16.4 b) 2.7 c) 0.16 d) 2.38

Prob 7: A mixture of methyl alcohol and nitrogen at a total pressure of 189 kPa contains 25 vol
% methyl alcohol. If the vapor pressure of methyl alcohol can be expressed by two
constant equation : $\log_{10} p = 7.846 - 1978.4/T$ where p is in kPa , T is in degrees K,
the dew point of the mixture (°C) is very nearly
  a) 320 b) 117 c) 47 d) 50

Prob 8: A vessel contains a liquid mixture of 50 % benzene and 50 % toluene by weight at 100 °C. The
vapor pressures at 100 °C are as follows (Assume Raoult's Law is followed.)
  Benzene C_6H_6 p = 1340 mm Hg Toluene C_6H_8 p = 560 mmHg
The average molecular weight of the vapor in contact with the solution is very nearly
  a) 82 b) 88 c) 78 d) 92

Prob 9: A solution of Na_2SO_4 in water is saturated at 50 °C. When a saturated solution of Na_2SO_4 is
cooled, crystals of Na_2SO_4 $10H_2O$ separate from the solution.

Temperature, °C	Solubility of Na_2SO_4, g/100g water
50	46.7
10	9

If 1000 kg of this solution is cooled to 10 °C, the percentage yield obtained is nearly
  a) 91 b) 90 c) 100 d) 80

Prob 10: A Carnot engine absorbs 1000 kJ at 827 C and rejects heat at 27 C.
The work done by the engine is closest to:
a) 1000 kJ　b) 730 kJ　c) 550 kJ　d) 827 kJ

Prob 11: How many degrees of freedom has a system of partially decomposed $CaCO_3$?
a) 1　b) 2　c) 3　d) 4

Prob 12 : What is the the molar specific volume of a gas in m^3/kgmol at 1.722 MPa and 99 °C,
compressibility factor = 0.87
a) 5.1　b) 2.1　c) 1.6　d) 3.5

Prob 13: A heat engine absorbs 1055 kJ at 427 °C. and rejects heat at 38 °C. The work done in kJ by
the engine if its efficiency is 50 % of the Carnot efficiency is closest to:
a) 496.6　b) 293.3　c) 1265　d) 320

Prob 14 : 80 kg of water at 95 °C are adiabatically mixed with 20 kg of cold water at 40 °C. The entropy
change for the process is closeset to:
a) 3.39 kJ/K　b) 5.0 kJ/K　c) - 3.39 kJ/K　d) - 5.0 kJ/K

Prob 15: The following data are available for the refrigerant HFC-134a :

T °C	30	40	50
P Bar	7.709	10.174	13.187
V_l m³/kg	8.431×10^{-4}	8.744×10^{-4}	9.056×10^{-4}
V_g m³/kg	0.02665	0.01998	0.01511

On the basis of above data, the heat of vaporization of HFC-134a at 40 °C (kJ/kg) is
closest to:
a) 85　b) 110　c) 171　d) 152

Prob 16: The two component system of phenol and O-cresol is found to obey Raoult's law.
At a temperature of 117 °C, the vapor pressures of phenol and O-cresol are 11.6 (P_1) and
8.76 (P_2) kPa respectively. The operating total pressure is (P_t) 10 kPa.

$$y_1 = \frac{P_1 x_1}{P_t} \qquad x_1 = \frac{P_t - P_2}{P_1 - P_2} \qquad \alpha_{12} = \left(\frac{y_1}{1 - y_1}\right)\left(\frac{1 - x_1}{x_1}\right)$$

The relative volatility of phenol relative to O-Cresol at these conditions is closest to
a) 1.28　b) 1.325　c) 1.16　d) 1.5

Prob. 17: A benzene -toluene feed (with 40 mol% benzene and 60 mol% toluene) to a distillation column is at a temperature of 20 °C. The molar specific heat of the feed is 159.2 kJ/(kmol.K). Molar latent heats of vaporization of benzene and toluene are 30813 and 33325 kJ/kmol respectively. The bubble point of the mixture is 95 °C. The slope of the q line is closest to:

    a) 0 b) 1 c) 3.7 d) -3.7

Prob 18: In absorption of NH_3 by water in a wetted wall column, the mass transfer coefficients in the gas and liquid phases were determined to be 0.6 kmol/(h.m².bar) and 40 kmol/(h.m².atm) respectively. The equilibrium relationship over the concentration range used in the experiment is given by y = 0.9 x. The overall mass transfer coefficient based on the liquid phase in $kgmol/h-m^3$ is

    a) 5.33   b) 0.533   c) 1.88   d) 0.0533

Prob 19: The following data pertain to a distillation column fractionating a feed mixture of 40 mol % benzene and 60 mol % toluene.
    Feed temperature = 20 °C (t_F)
    Distillate D: flow = 42 kmol/h, composition: Benzene 97 mol %
    Bottoms product B : flow = 63 kmol/h, composition : Toluene 98 mol %
    Bottom product temperature = 109.4 °C (t_B)
    Distillate temperature = 82.2 °C (t_D)
    External reflux ratio R = 3.7
    Molar Specific heat of feed bottoms and distillate = 159.2 kJ/(kmol.K)
    Latent heats of vaporization: Benzne 30813 kJ/kgmol. (λ_1)
                              Toluene 33325 kJ/kgmol. (λ_2)
    Enthalpy of distillate relative to feed = $Dc_p(t_D-t_F)$
    Enthalpy of bottoms relative to feed = $Bc_p(t_B-t_F)$
    Condenser duty = D(R+1) λ_{AV}
        $\lambda_{AV} = x_1 \lambda_1 + x_2 \lambda_2$
    Assume total condenser

The reboiler duty in units of kW is closest to:
    a) 264 b) 2058 c) 380 d) 1700

Prob 20: In absorption of NH_3, the mass transfer coefficient in the gas phase was determined to be 0.59 kg-mol/(h.m²) and that in the liquid phase was found to be 40 kg-mol/(h. m²). The slope of the equilibrium line in the range of experiments was 0.9. The statement that correctly applies to mass transfer in this system is
    a) The liquid phase controls the mass transfer in this system.
    b) The gas phase controls the mass transfer in this system.
    c) Both the gas and liquid resistances are equally significant.
    d) Gas phase controls the mass transfer, therefore gas phase resistance is negligible.

Prob 21: Acetone is to be absorbed from an air stream containing 3 mole % acetone in an existing packed tower of 1.83 m diameter. The gas flow rate will be 212.5 m³/min at 30 °C and 1.2 bar. The overall mass transfer coefficient for acetone absorption in water may be taken as 230 kg-mol/h.m³.(mol fraction). Then the overall height (m) of a gas transfer unit (HTU) is closest to:
    a) 0.8 b) 1.0 c) 0.6 d) 1.16

Prob 22: A reacts at 25 ^{0}C according to the following reactions

$$A + B \xrightarrow{k_1} 2R + S \qquad \Delta H_R = -125.52 \; kJ/gmol \; A$$
$$A + B \xrightarrow{k_2} P \qquad \Delta H_R = -83.08 \; kJ/gmol \; A$$

The reaction rate constants for the two reactions are as follows

$$k_1 \quad 1.2 \; \frac{liters}{gmol.\min} \qquad k_2 = 0.3 \; \frac{liters}{gmol.\min}$$

If 2 mols of A react, the total heat of reaction in kJ is

     a) 234.3   b) - 234.3   c) - 251   d) - 209

Prob 23: A batch reactor is used to convert a liquid reactant A. The reaction follows second order kinetics. The time in hours required to convert 60 % of A if the initial concentration of A is 0.8 gmol/L and k = 4 L/(g-mol.h) is

     a) 4.7   b)   0.94   c)   0.47   d)   0.5

Prob 24: The following data are obtained at 200 ^{0}C for the reaction 2A → 2B + C . Initially A is pure. The concentration of A changes as follows

Time Min	0	200	300	500
Concentration of A gmol/L	0.02	0.01585	0.01435	0.0128

The order of the reaction is

     a) 2   b)   1   c)   1.5   d)   3

Prob 25: A homogeneous gas phase reaction A → 2R has a reaction rate at 250 ^{0}C and 2 bar as

$$-r_A = 10^{-2} C_A \; (g-mol/liter.s)$$

The space-time in minutes needed for 90 % conversion of A in a plug-flow reactor operating at 250 ^{0}C and 2 bar if the feed to the reactor is 50% A and 50% inerts is

     a) 300   b)   30   c)   60   d) 5

Prob 26-27: A liquid phase reaction A → R is carried out in a series of three batch reactors of equal size. The reaction rate constant k is 0.066 min^{-1}. Overall conversion is 90 %. The feed rate is 10 liters/min and the feed contains only A in concentration of 1g-mol/L.

Prob 26: The overall space-time in minutes for the three reactor system

     a) 5.24   b)   524   c)   52.4   d)   17.5

Prob 27: The concentration from the second reactor is

     a) 0.2148   b)   0.1   c)   0.35   d)   0.5

Prob 28
The weight of a metal sphere is 1000 N at a place where the local acceleration of gravity g is 10 m/s^2. The weight of the metal sphere on the surface of the moon where the acceleration due to gravity is 1.67 m/s^2 is closest to:

        a) 10N   b)   170N   c) 250N   d)   500N

Prob 29
A barometer reads 800 mmHg. The atmospheric pressure at this location in millibars, if the specific gravity of mercury is 13.6, and g has the value of 9.807 m/s^2 is closest to:

        a) 953 mBar   b) 990 mBar   c) 1030 mBar   d) 1070 mBar

Prob 30   For a fractionating column separating a mixture of benzene and toluene, the slope of q line was - 0.5. 12 theoretical stages are stepped off on the McCabe-Thiele diagram. The operating lines for the rectifying and stripping sections intersect on q line between 7th and 8th theoretical stages measured from the top of the column. The overall efficiency of distillation is 70 %. Actual feed tray location will be on the tray:

        a) 7   b) 8   c) 10 d) 11

Prob 31
The fixed cost per day of a plant producing a certain type of chemical is $10,000/day. The selling price of the chemical is $200/kg. The variable cost may expressed as 50+0.1P ($/kg) where P = production rate in kg/day. Find the break-even capacity of the plant.

    (a)70   (b)1430   (c) 70 or 1430   (d) 100

Prob 32   The feed to a binary distillation column is 25 mol % vapor and the balance liquid. The slope of the q line with respect to the positive direction of the X axis on the x-y diagram is closest to:

        a) 0.67   b) 3.0   c) -3.0 d) - 0.67

Prob 33 The total annual venture cost for insulating an oven may be expressed by:

$$Total\ annual\ venture\ cost\ =\ \frac{870}{0.25\ +\ 2.78T}\ +\ 17.25T$$

where T = insulation thickness . Find the optimum economic insulation thickness.
  (a)2   (b)3   (c)1   (d)4

Prob 34
The walls of a brick-lined house consists of the following layers materials:
 (a)Brick layer, 0.1 m thick, k = 0.8 W/m.K
 (b)Rock-wool insulation, 0.0762 m thick, k = 0.065 W/m.K
 (c)Gypsum plaster board, 0.0375 m thick, k = 0.5 W/m.K
If the inside of the house is maintained at 295K, estimate the heat loss by conduction through the walls of area 200 m^2 when the outside temperature is 265K.
  (a) 4400 W   (b) 2200 W   (c) 8400 J/s   (d) Insufficient information

Prob 35
The overall heat transfer coefficient(U) of a clean shell and tube heat exchanger with negligible tube wall resistance, and equal film coefficients is 100 W/m²K. If the shell side coefficient remains constant, but the tube side coefficient is lowered by 10%, find the ratio $U_{final}/U_{original}$.
  (a)0.80 (b)0.95 (c)0.90 (d) 1.10

Prob 36
The following information on the terminal temperatures are available for a (2-4) shell-and-tube countercurrent heat exchanger. Estimate the correction factor for LMTD.

	Shell Side	Tube Side
Temperature in K	500	200
Temperature out K	300	290

   (a)1.1 (b)0.75 (c)0.98 (d)0.8

Prob 37
A batch is to be cooled from 5°C to -17°C in a stirred tank jacketed vessel with an isothermal cooling medium vaporizing at -26°C. The jacket has heat transfer surface of 10 m^2 with an overall heat transfer coefficient of 850 W/m²K. What will be the peak instantaneous cooling load in W, ignoring the heat input from the stirrer, and jacket area is fully wetted by the batch?
  (a)120,000   (b)1,120,000   (c) 163,500   (d) 263,500

Prob 38
A double-pane window assembly consists of two glass plates 0.0025 m thick(k_{glass} = 0.79 W/m.K) with 0.01 m air gap between the plates (k_{air} = 0.026 W/m.K). The temperature difference between inside and outside is 25 K. The inside and outside convection coefficients are 6 W/m².K and 25 W/m².K respectively.The heat loss through a double-pane glass window in W/m² is closest to:

  a) 20 W/m²   b) 40 W/m²   c) 60 W/m²   d) 80 W/m²

Prob 39
A black asphalt street absorbs total solar radiant energy at the rate of 1104 W/m² during summer. If the convection heat transfer coefficient between the street and air is 15.9 W/m²K, and the ambient temperature is 305K, then the equilibrium temperature of the street is closest to:

  a) 320K   b) 351K   c) 400K   d) 410K

Prob 40 A Newtonian fluid is flowing in a pipe at a Reynolds number of 1500. If the viscosity is lowered to 50% and density to 60% of their original values by heating, what will be the effect of pressure drop for the same flow rate and pipe dimensions?

  (a)50% of original (b) 60% of original (c)25% of original (d)30% of original

Prob 41 Water is flowing under fully turbulent flow in a pipe. If the diameter of the pipe is lowered by 10%, what will be the effect on pressure drop for the same flow rate?

  (a)Increase by 10% (b) Increase by 21% (c) Increase by 59.05% (d)Increase by 69.35%

Prob 42 Temperature rise in a heat sensitive organic liquid due to skin friction is to be limited to 10 K. What is the maximum allowable frictional pressure drop if the specific heat and density of the liquid are 2000 J/kg.K and 500 kg/m³ respectively?

  (a)1E4 Pa (b) 1E7 Pa (c) 1E5 Pa (d)1E2 Pa

Prob 43 Estimate an approximate pressure drop in psi/100ft for 500 SCFM of air compressed to 100 psig at 60 °F flowing through a 3.068 "ID pipe with roughness factor of 0.00015 ft.

  (a)0.09 (b) 0.19 (c) 0.29 (d) 0.39

Prob 44 Given:The density of fluid is 1.1 x10³kg/m³. The barometric pressure is 100kN/m², the local acceleration due to gravity is 9.81 m/s². The absolute static pressure at the bottom of a column of fluid of height 6m., open to atmosphere is closest to:

  a) 150 kN/m²   b) 165 kN/m²   c) 65 kN/m²   d) 70 kN/m²

Prob 45

In an effort to increase the flow rate of a Newtonian liquid, the supply pressure is increased by 40% in a pipe. Assuming fully turbulent flow before and after the increase of pressure, and no change in density or viscosity, what will be the % increase of flow?

      a) 12   b) 18   c) 24   d) 30

Prob 46   A centrifugal fan with a blade diameter of 0.8m operates at 20rps to deliver 7.3 m³/s of air. What would be the capacity in m³/s of a geometrically similar fan with 1.23m impeller blade operating at 30rps?

      (a) 50   (b) 19   (c) 38   (d) 25

Prob 47   Feed, distillate and bottoms product flows and their compositions for a fractionating column are given in the following table (the components are listed in the decreasing order of their K values):

Component	FEED mol %	distillate mol %	Bottoms product mol %
Benzene	15.89	18.69	0.0
Toluene	68.79	80.89	0.04
Ethyl benzene	0.7	0.04	4.44
p-Xylene	11.63	0.34	75.77
m-Xylene	1.41	0.03	9.25
o-Xylene	0.31		2.08
Ethyl toluene	0.98		6.55
Heavies	0.28		1.87
Flow (kgmol/h)	2222.7	1939.05	283.64

Based on the above data , the heavy key component is

      a) Toluene   b) Ethyl benzene   c) p-Xylene   d) Ethyl toluene.

Prob 48 A shell and tube condenser with saturated fluid condensing in the shell side and liquid flowing through the tube side in turbulent flow without change in phase with negligible tube wall resistance has 1000 W/m²K film coefficient in the shell side and 125 W/m²K in the tube side. Estimate the overall coefficient if the tube side velocity is doubled, without changing other physical properties.

      (a) 179 (b) 150 (c) 250 (d) 161.5

Prob 49   What is the upper flammability limit and lower flammability of a mixture of gases whose composition and individual UFL and LFL are given below.

Component	Volume %	LFL %	UFL %
heptane	1	1.1	6.7
hexane	1	1.1	7.5
hydrogen	3	4	75
isopropyl alcohol	1	2	12
Total combustible	6		
air	94		

(a) LFL=1.95%, UFL= 14.7%   (b) LFL= 2.7%, UFL = 41.8% (c)LFL= 5.3%. UFL = 28.9%
(d)LFL = 2.8%, UFL= 37.43%

Prob 50 The time to reduce the concentration of a toxicant from 2.5% by volume to 1 ppm by volume from a vessel of 5000 m^3 in capacity by sweeping an inert at the rate of 3000 m^3 through the vessel is closest to:. (Assume a non-ideal mixing factor of 0.2)
(a) 17 minutes (b) 25 minutes (c)85 minutes (d) 50 minutes

Prob 51 For the situation given in prob 50, the number of changes of container-volumes for the above duty is closest to:
(a) 10 (b)20 (c)40 (d)51

Prob 52
Calculate the pH of an aqueous waste containing HCl(0.01 molar) and KCl(0.09 molar) at 25°C.
(a)2.5 (b)1.0 (c) 3.0 (d) 2.1

Prob 53

100 kg of saturated liquid water at 350 kPa contained in a closed vessel are heated
until 80 % of water is vaporized. The amount of heat added to the system is closest to?
(a) 58100 kJ   (b) 93000 kJ   (c) 127900 kJ   (d) 172000 kJ

Prob 54

The stack analysis of an ammonia scrubber shows ammonia concentration
of 180 ppm(v/v) at 30°C and 1.064 Bar. The concentration of ammonia
in the stack in mg/m³ is closest to:
(a) 90 mg/m³   (b) 130 mg/m³   (c) 170 mg/m³   (d) 260 mg/m³

Prob 55

Cyanuric Chloride has a flash point higher than 374°F. A 75000 gallons storage tank to store cyanuric chloride liquid is installed so that the tank is not within the diked area or drainage path of Class I or Class II liquids. Is it necessary to size the emergency relief venting of the tank for fire exposure according to NFPA?

        a) yes   b) no

Prob 56

Which of the following substance(s) is (are) not considered a VOC:

        a) Carbon Dioxide b) Carbon disulfide c) Carbon Tetrachloride   d) Benzene

Prob 57

Which of the following emissions is (are) not a fugitive emission(s):

        a) Vapor emission from the top vent of a scrubber
        b) Emission from a pump seal
        c) Leakage from a relief valve
        d) Leakage from a compressor

Prob 58. The following polynomial obviously has three roots:

$$P(x) = x^3 + 4x^2 + 6x + 4$$

    Is $x = -1$ one of the exact roots?

a) No because its less than the last coefficient
b) Yes, because there is a remainder when using synthetic division.
c) No, because there is a remainder when using synthetic division.
d) Yes, there is no remainder.

Prob 59. For above problem if a second trial root to be found (from equation 13-4) what is this next trial division?

a) 1
b) 2
c) -1
d) -2

Prob 60. For a differential equation: $dx/dt + 5x = 1$, $x(0)=0$; What is the value of x suggested by the Euler's method after first iteration after initial condition?

a) 0.1
b) 0.2
c) 0.3
d) 0.4

END of Examination. Stop

Solutions to Sample Exam Problems :

Prob 1-3:

Atomic and molecular weights $Fe = 56$ $H_2 = 2$ $H_2O = 18$ $Fe_3O_4 = 232$

Stoichiometric relation $3Fe + 4H_2O \rightarrow Fe_3O_4 + 4H_2$

                         168   72   232   8

Basis   100 kg hydrogen produced

Prob 1:

Fe reuired $= (168/8)(100) = 2100$ kg   **Answer is c)**

Prob 2:

Fe_3O_4 produced $= (232/8)(100) = 2900$ kg   **Answer is b)**

Prob 3:

$H_2 = 100/2 = 50$ kmol

1 kmol of an ideal gas occupies 22.4 m^3 volume at standard conditions

Therefore, volume of 50 kmols of $H_2 = 50 \times 22.4 = 1120$ m^3 at standard conditions

                                            Answer is b)

Prob 4:

Two data points are available or given. Therefore, use two constant equation to correlate the vapor pressure data.

The two constant equation is $\ln p = A + B/T$ p in kPa T= K

VP at 0 °C = 3.22 kPa,   VP at 160 °C = 845.8 kPa

By substitution, get two equations as follows

      $\ln 3.22 = A + B/(273 +0) = A + B/273$

      $\ln 845.8 = A + B/(273 + 160) = A + B/433$

By simplification ,

$273A + B = 319.24$ and $433A + B = 2918.54$

solving simultaneously, $A = 16.246$ and $B = -4116$

vapor pressure at 100 °C is given by $\ln p = 16.246 - 4116/(273+100)$

$\ln p = 5.211$ Therefore, $p = 183.3$ kPa   **Answer is a)**

Prob 5-6:

Basis 100 kmols of gas

Reactions are $CH_4 + 2O_2 = CO_2 + 2H_2O$

                   $C_2H_6 + 3.5 O_2 = 2 CO_2 + 3 H_2O$

Assume air composition $O_2 = 21\%$ by vol. $N_2 = 79\%$ by vol.

Prepare a table showing reactants and products as follows

Component	kmols feed	O$_2$ consumed kmols (thoretical)	O$_2$ kmol (Excess)	N$_2$ kmol in air	products kmol/100 kmol
CH$_4$	85.0	170			
C$_2$H$_6$	10.5	36.5			
N2	4.5			893.36	897.86
O$_2$		30.975	30.975		
CO$_2$					106
H$_2$O					201.5

Prob 5:

Products per 10 kmol of fuel gas burned

$N_2 = 89.79$ kmol $O_2 = 3.10$ kmol $CO_2 = 10.6$ kmol $H_2O = 20.15$ kmol

Therefore, total kmols/10 kmol of gas burned = 123.64 kmol   **Answer is d)**

Prob 6:

Water vapor in the gas/10 kmol = 20.15 kmol

Mol fraction of water = 20.15/123.64 ≅ 0.163

Partial pressure of water vapor in product gas = 0.163x100.6 = 16.4 kPa

Answer is a)

Prob 7:

Dew point is the temperature at which vapor pressure of methyl alcohol in the mixture will be equal to its vapor pressure at that temperature.

Total pressure = 189 kPa

Mol fraction of MeOH = 0.25 (volume fraction = mol fraction, ideal behavior assumed)

Partial pressure of MeOH = 0.25 x 189 = 47.25 kPa

Substituting in the vapor pressure equation gives

$$\log 47.25 = 7.846 - 1978.4/T$$

$$1.6744 = 7.846 - 1978.4/T$$

$$-6.1716 = -1978.4/T \text{ or } T = \frac{-1978.4}{-6.1716} = 320.6 \text{ K}$$

Therefore, t = 320.6 - 273.= 47.6 °C **Answer is c)**

Prob 8: Basis 100 kg of mixture. Prepare a table as follows

component	wt%	wt. kg	MW	kg-mols	x_i (liquid)	P_i kPa	p_i (vapor)	y_i
Benzene	50	50	78	0.6410	0.5412	178.6	96.7	0.7387
Toluene	50	50	92	0.5435	0.4588	74.6	34.2	0.2613
				1.1845	1.0000		982.1	1.0000

Therefore, vapor composition : benzene 73.87 mol% and toluene 26.13 mol%

Average molecular wt. of vapor = 0.7387(78) + 0.2613(92) = 81.66 **Answer is a)**

Prob 9: At 50 °C, solubility of Na_2SO_4 = 46.7 g/100 g of water. i.e. 31.83 % by weight

at 10 °C, solubility of Na_2SO_4 = 9 g /100 g of water i.e. 8.26 % by weight

Let x be the amount of $Na_2SO_4 \cdot 10H_2O$ formed in kg

By material balance on Na_2SO_4, 1000(.3183) = (142/322) (x) + (1000 - x) (0.0826)

which on solution gives x = 657.6 kg of Na_2SO_4 $10H_2O$

Na_2SO_4 in original solution = 318.3 kg and Na_2SO_4 in crystals = (0.441)(657.6) = 290 kg

% yield = (290/318.3)x100 = 91.1 % **Answer is a)**

Prob 10:

$$T_H = 827 + 273 = 1100 \text{ K} \qquad T_C = 27 + 273 = 300 \text{ K}$$

Work done = heat absorbed x (Carnot efficiency)

$$= 1000 \text{ x } \frac{T_H - T_C}{T_H} = 1000 \text{ x } \frac{1100 - 300}{1100} = 727.3 \text{ kJ.} \quad \textbf{Answer is b)}$$

Prob 11

The decomposition of $CaCO_3$ can be represented by the equation :

$$CaCO_3(s) \rightarrow CaO(s) + CO_2(g)$$

Number of components = 3, Number of reactions = 1, number of phases = 3 (2 solid and 1 gas)

By phase rule, the degrees of freedom = N - r + 2 - P = 3 - 1 + 2 - 3 = 1 **Answer is a)**

Prob 12:

T = 99 + 273 = 372 K

$$v = \frac{ZRT}{P} = \frac{0.87 \times 0.0008314 \times 372}{1.722} = 1.563 \text{ m}^3\text{/kg-mol} \qquad \textbf{Answer is c)}$$

Prob 13:

$T_1 = 427 + 273 = 700$ K $T_2 = 38 + 273 = 311$ K

Carnot effciency = (700 - 311) / 700 = 0.556

Actual efficiency = 0.556 x 0.5 = 0 .278

Work done = 0.278 x 1055= 293.3 kJ.   **Answer is b)**

Prob 14 :

Assume specific heat of water constant and equal to 4.184 kJ/kg-K

By energy balance,   4.184x 80 (95-t) = 20(t -40) x 4.184

Solving for t, gives t = (95x80 + 20x40)/100 = 84 ^{0}C = 357 K

For hot water, $\Delta S_H = 80 \int_{368}^{357} \frac{C_p dT}{T} = 80(4.184) \ln \frac{T_2}{T_1} = 80(4.184) \ln \frac{357}{368}$ = - 10.16 kJ/K

For cold water, $\Delta S_C = 20 \int_{313}^{368} \frac{C_p dT}{T} = 20(4.184) \ln \frac{368}{313} = +13.55$ kcal/K

Total entropy change for the mixing process is

$\Delta S_H + \Delta S_C = -10.16 + 13.55 = +3.39$ kJ/K   **Answer is a)**

Prob 15:

Assume ΔH_V = constant over the small temperature range.

Then $\frac{\Delta P}{\Delta T} = \frac{\Delta H}{T \Delta V}$ Bar/K

$\Delta H_V =$ T $(\Delta V)(\Delta P/\Delta T) = (40 + 273)(0.01998 - 8.744 \times 10^{-4})(0.2739)$

= 1.7074 m^3 Bar/kg

= 1.7074 (m^3 Bar/kg) / (10^{-2}m^3 Bar/kJ)

= 170.74 kJ/kg   **Answer is c)**

Prob 16:

Let x_p = mol fraction phenol in liquid y_p = mol fraction phenol in vapor.

Likewise x_{oc} and y_{oc} are liquid and vapor compositions of o-Cresol

P_p and P_{oc} are vapor pressures of pure phenol and o-Cresol respectively.

By Raoult's law, $y_p = \frac{x_p P_p}{P_T} = \frac{x_p P_p}{10}$ $y_{oc} = \frac{x_{oc} P_{oc}}{P_T} = \frac{x_{oc} P_{oc}}{10}$

Also $x_{oc} = 1 - x_p$

By Dalton's relation, $x_p P_p + (1 - x_p) P_{oc} = 10$

from which $x_p = \frac{10 - P_{oc}}{P_p - P_{oc}}$

For given data of the problem,

$x_p = \frac{10 - 8.76}{11.6 - 8.76} = 0.43$

$y_p = 0.437 \times 11.6 / 10 = 0.507$

$\alpha_{phenol} = \frac{y_p x_{oc}}{y_{oc} x_p}$ $= \frac{y_p (1 - x_p)}{(1 - y_p)(x_p)}$

$= \frac{0.507(1 - 0.437)}{(1 - 0.507)((0.437)} = 1.32$

Answer is b)

Prob 17:   Latent heat of vaporization of the mixture

= 0.4(30813) + 0.6(33325) = 32320 kJ/kmol of feed.

Sensible heat to raise the temperature of feed to bubble point

= 159.2 (95-20) = 11940 kJ/kmol

Then $q = \frac{32320 + 11940}{32320} = 1.3$

and the slope of q line = $\frac{1.37}{1.37-1}$ = 3.

Answer is c)

Prob 18:

Based on the liquid phase, the overall mass transfer coefficient is given by

$$\frac{1}{K_x} = \frac{1}{k_x} + \frac{1}{mk_y} = \frac{1}{40} + \frac{1}{0.9\times(0.6)} = 1.8769$$

Therefore, K_x = (1/1.6769) = 0.533 kmol/h.m² **Answer is b)**

Prob 19:

Heat balance on the column gives $Q_R = Q_c + Q_{SB} + Q_{SD}$
Assume specific heats of feed, distillate and bottoms product same.
$Q_{SD} = 159.2(82.2 - 20)(42) = 415894$ kJ/h
$Q_{SB} = 159.2(109.4 - 20)(63) = 896646$ kJ/h
Calculation of Q_c:
vapor to condenser = D(R + 1) = 42(3.7 + 1) = 197.4 kmol/h
λ = 0.97(30813) + 0.03(33325) = 30888 kJ/kmol
Q_c = 30888 197.4 = 6097362 kJ/h
Then the reboiler duty Q_B = 415894 + 896646 + 6097362 = 7409902 kJ/h
= 7409902/3600 = 2058 kJ/s = 2058 kW

Answer is b)

Prob 20:

Calculate the resistances in each phase based on gas phase.

$$\frac{1}{K_y} = \frac{1}{k_y} + \frac{m}{k_x} = \frac{1}{0.59} + \frac{0.9}{40} = 1.695 + 0.0225 = 1.7175$$

The liquid phase resistance is 0.0225 compared to 1.695 in the gas phase.
Therefore in this system , the gas phase controls the mass transfer process. **Answer is b)**

Prob 21:

Cross sectional area of tower = 0.785 (1.83)² = 2.63 m²

Gas rate = $\frac{212.5\times60}{22.4} \times \frac{273}{303} \times \frac{1.22}{1.013} = 617.64$ kg – mol/h

G = $\frac{617.64}{2.63} = 234$ kgmol/h –m²

Therefore, HTU$_G$= (G/K$_y$a) = 234/230 = 1.017 m **Answer is b)**

Prob 22:

For the first reaction, $-r_{A1} = k_1 C_A C_B$

For the second reaction, $-r_{A2} = k_2 C_A C_B$

Dividing first equation by the second, $\dfrac{r_{A1}}{r_{A2}} = \dfrac{k_1}{k_2} = \dfrac{1.2}{0.3} = 4.0$

Thus the ratio of mols of A consumed in reaction 1 to mols of A consumed in the second reaction is 4:1.
Then mols of A reacted in first reaction $= 2\,(4/5) = 1.6$ mols
Also mols of A reacted in second reaction $= 2(1/5) = 0.4$ mols
Total heat of reaction $= 1.6(-125.52) + 0.4(-83.08) = -234.3/2$ mols of A

Answer is b)

Prob 23:

For a reaction following second order kinetics, $-r_A = -\dfrac{dC_A}{dt} = kC_A^2$
In terms of conversion X_A, the rate equation becomes

$$-r_A = C_{AO}\frac{dX_A}{dt} = kC_{AO}^2\left(1 - X_A\right)^2 \quad \text{Then} \quad C_{AO}kt = \int_0^{X_A} \frac{dX_A}{\left(1 - X_A\right)^2}$$

Solution of the equation is $\qquad C_{AO}kt = \dfrac{X_A}{1 - X_A}$

For a batch reactor, $\qquad t = \dfrac{X_{A'}\left(1 - X_A\right)}{C_{AO}\,k} = \dfrac{0.6/(1 - 0.6)}{0.8 \times 4} = 0.47\ hr$

Answer is c)

Prob 24:

First assume the reaction is second order. The rate equation is $-\dfrac{dC_A}{dt} = kC_A^2$ or
$-\dfrac{dC_A}{C_A^2} = k\,dt$ which on integration with boundary condition, $C_A = C_{AO}$ at $t = 0$

yields the solution

$$\frac{1}{C_A} - \frac{1}{C_{AO}} = kt$$

Thus a plot $\dfrac{1}{C_A}$ of versus t will be a straight line. To solve the problem without a graph, calcu late k values for each run. If k values thus calculated are the same, the assumed order of the reaction is correct. The calculations are given below

time (min)	C_A	$1/C_A$	$\dfrac{1}{C_A} - \dfrac{1}{C_{AO}}$	$k = \left(\dfrac{1}{C_A} - \dfrac{1}{C_{AO}}\right)/t$
0	0.02	50.00		
200	0.01585	63.09	13.09	0.0655
300	0.01435	69.68	19.68	0.0656
500	0.01218	82.78	32.78	0.0656

k values calculated are equal, therefore assumed order of the reaction is correct.

Answer is a)

Prob 25:

Change of volume :		Initial mols	Final mols
	A	1	0
	R	0	2
	I	1	1
		2	3

Therefore ,
$$\epsilon_A = \frac{3-2}{2} = 0.5$$

And the relation between conversion and concentration is :

$$\frac{C_A}{C_{AO}} = \frac{1-X_A}{1+\epsilon_A X_A} = \frac{1-X_A}{1+0.5X_A}$$

For a plug-flow reactor, $\quad \tau = C_{AO}\int_0^{X_A}\frac{dX_A}{-r_A}$

$$= \quad C_{AO}\int_0^{X_A}\frac{dX_A}{10^{-2}C_A} = C_{AO}\int_0^{X_A}\frac{dX_A}{10^{-2}C_{Ao}(1-X_A)'(1+0.5X_A)}$$

Solution of which is $\quad 10^{-2}\tau = -(1+\varepsilon_A)\ln(1-X_A) - \epsilon_A X_A$

$$= -1.5\ln(1-X_A) - 0.5X_A$$

$$\tau = 10^2\left[-1.5\ln(1-0.9)-0.5\times0.9\right]$$
$$= 300.4\ s = 5\ min. \qquad\qquad \textbf{Answer is d)}$$

Prob 26-27:
Prob 26:

For a system of three reactors, $\quad \tau n = 3\tau = \frac{n}{k}\left[\left(\frac{C_{AO}}{C_{An}}\right)^{\frac{1}{n}} - 1\right]$

$$= \frac{3}{0.066}\left[\left(\frac{1}{0.1}\right)^{\frac{1}{3}} - 1\right]$$

$$= 52.4\ min \qquad\qquad \textbf{Answer is c)}$$

Prob 27:

Concentrations of A in exit stream from each reactor

Since all reactors are of equal size, $\quad \frac{C_{Ai-1}}{C_{Ai}} = 1 + k\tau \qquad$ for first order reaction

for each reactor $= 52.4/3 \cong 17.5\ min$

Then $\qquad C_{A1} = \frac{C_{Ai-1}}{1+k\tau} = 1/(1+0.066\times17.5) = 0.4635$

$$C_{A2} = 0.4635(1+0.066\times17.5) = 0.2148$$

$$C_{A3} = 0.2148(1+(0.066\times17.5)) = 0.1 \qquad\qquad \textbf{Answer is a)}$$

Prob 28

By Newton's law, F=ma=mg since a=g

or since weight and force ar the same, weight = mg

Hence, m=weight/g = 1000 N/(10.0 m/s^2)

$\qquad$ =100.0 Nm/s^2 = 100 (1 kg/m-s^2) = 100 kg

Mass of the metal sphere will remain the same regardless of its location. However, its weight will change because of a change in acceleration due to gravity. Since the weight and the gravitational force will be the same on the surface of the moon, one can write

$\qquad$ weight = F_{moon} = mg = 100 kg x 1.67 m/s^2=167 N $\qquad$ ANSWER is b).

Prob 29

It can be shown that the following relation holds for a column of mercury in a barometric tube

$\qquad$ $\Delta P = P_A - P_V = -\rho g \, \Delta Z$

where $\quad$ PA = Atmospheric pressure

$\qquad$ PV = Pressure exerted on the surface of the liquid column in the barometric tube

$\qquad$ = 0 for a mercury barometer.

$\qquad$ Z $\;$ = height of mercury column in the barometric tube.

For mercury, the specific gravity is 13.6, hence its density is : ρ = 13.6 g/cm^3 = 13.6 x 10^3 kg/cm^3

$\qquad$ ΔZ = 800 mm/(1000mm) = 0.8 m

$$P_A = (13.6 \times 10^3 \, kg/m^3) \times (9.807 m/s^2) \times (0.8 m)$$

$$= 1.067 \times 10^5 \, \frac{kg}{ms^2}$$

$$= 1.067 \times 10^5 \, Pa \quad \text{since } 1 \text{ Pa} = 1 \frac{kg}{ms^2}$$

$$= 1.067 \times 10^5 \, Pa / (10^5 \, Pa / Bar)$$

$$= 1.067 \;\; Bar \; = \; 1067 \; mBar$$

$\qquad\qquad\qquad\qquad\qquad\qquad$ ANSWER is d).

A simpler and quicker way to solve this problem is to use the conversion factors from the table.

Prob 30 No. of theoretical stages in the column = 11

$\quad$ Feed is on 7th theoretical stage measured from the top.

$\quad$ Or 7/0.7 = 10 th actual tray in the column.

$\qquad\qquad\qquad\qquad\qquad\qquad$ **Answer is c)**

Prob 31

$$Q_B = \frac{F}{s-c}$$

F = $10,000/day
s = $200/kg
c = 50 + 0.1P. At break-even point, P = break-even capacity = Q_B kg/day.
 Therefore, at break-even capacity, c = 50 + 0.1Q_B.

Substituting in the above equation:

$$Q_B = \frac{10000}{200 - (50 + 0.1Q_B)} = \frac{10000}{150 - 0.1Q_B}$$

$$Q_B^2 - 1500Q_B + 100000 = 0$$

Solving for Q_B, Q_B = 70 or 1430

ANSWER is c)

Prob 32 Since vapor is 25 mol % , the liquid fraction of the feed is 75 mol %.

Therefore the slope of the q line = $\frac{0.75}{0.75-1} = -3.0$

Answer is c)

Prob 33

$$V_c = \frac{870}{0.25 + 2.78T} + 17.25T$$

$$\frac{dV_c}{dT} = - \frac{870 \times 2.78}{(0.25 + 2.78T)^2} + 17.25 = 0$$

Hence, T = 4.2   **ANSWER is d)**

Prob 34

From equation (7-3):

$$q_k = \frac{T_o - T_3}{X_1/k_1A + X_2/k_2A + X_3/k_3A}$$

$$= \frac{A(T_o - T_3)}{X_1/k_1 + X_2/k_2 + X_3/k_3}$$

$$= \frac{200(295-265)}{0.1/0.8 + 0.0762/0.065 + 0.0375/0.5}$$

$$= 4372.2 \text{ W}$$

ANSWER is a)

Prob 35

Originally:

$$U_{o1} = 100 = h_o h_i/(h_o + h_i) = h_o/2, \text{ since } h_o = h_i$$

$$h_o = h_i = 2U_{o1} = 200$$

Finally:

$h_o = 200, h_i = 0.9 \times 200 = 180$
$U_{o2} = 200 \times 180/(200+180) = 94.74$

$U_{o2}/U_{o1} = 0.9474$ **ANSWER is b)**

Prob 36

$T_1(500)$-----------------$>T_2(300)$
$t_2(290)<$-----------------$t_1(200)$

From Perry[4]

$R = (T_1-T_2)/(t_2-t_1) = 200/90 = 1.11$
$S = (t_2-t_1)/(T_1-t_1) = 90/300 = 0.3$
$F^T \simeq 0.98$ **ANSWER is c)**

Prob 37
From equation (7-37):
$-Q_{\theta=0} = UA(T_1-t_1) = 850 \times 10[5-(-26)] = 263,000$ W   Answer d).

Prob 38
From equation (7-6a):

$1/U = 1/h_i + \Sigma X_i/k_iA_i + 1/h_o$
$= 1/6 + 2 \times 0.0025/0.79 \times 1 + 0.01/0.026 \times 1 + 1/25$

           2 glass panes   air gap

$= 0.5976$ m^2.K/W
$U = 1.6733$ W/m^2.K
From equation (7-6b):
$q = UA\Delta T = 1.6733 \times 1 \times 25 = 41.8325$ W/m^2 ANSWER is b).

Prob 39
From equation (7-14), taking ε_1 as 1 for asphalt:
$q_{r12}/A_1 = 5.67[(T_{street}/100)^4 - (T_{air}/100)^4] + h_{c,street}(T_{street}-T_{air}) = 1104$
$5.67[(T_{street}/100)^4 - (305/100)^4] + 15.9(T_{street}- 305) = 1104$
$5.67(T_{street}/100)^4 + 15.9(T_{street}) - 6444.2 = 0$
By trial and error: $T_{street} \simeq 351$ K   ANSWER is b).

Prob 40
Check N_{Re} for turbulent flow. For the same pipe and same flow rate, $N_{Re} \propto \rho/\mu$.

$N_{Re2} = N_{Re1}(\rho_2\mu_1/\rho_1\mu_2) = 1500 \times 0.6 \times 1/(1 \times 0.5) = 1800 < 2100$. Flow is still laminar. From equation (8-7): $\Delta P_2/\Delta P_1 = \mu_2/\mu_1 = 0.5/1 = 0.5$.
Therefore, the pressure drop will be reduced by 50%.   ANSWER is a).

Prob 41
From equation (8-15), $\Delta P \propto 1/d^5$. Therefore, $\Delta P_2/\Delta P_1 = (d_1/d_2)^5 = (1/0.9)^5 = 1.6935$. Hence, pressure drop will increase by 69.35%. ANSWER is d).

Prob 42

From equation (8-33), $\Delta P = \Delta T \rho \eta C_p = \dfrac{10\ K}{} \times \dfrac{500\ kg}{m^3} \times \dfrac{N.m}{J} \times \dfrac{2000J}{kg.K}$
$= 1 \times 10^7$ Pa.   ANSWER is b).

Prob 43
Assume fully turbulent flow. From equation (8-16):

$$f = \frac{1}{16[\log(3.7 \times 3.068/12 \times 0.000150)]^2}$$
$$= 0.00432$$

$\rho_{standard}$ = PM/RT = $14.7 \times 29/(10.73 \times 520)$ = 0.0764 lb/ft^3
Air mass flow rate = $500 \times 0.0764 \times 60$ = 2292 lb/h
ρ_{flow} = PM/RT = $114.7 \times 29/(10.73 \times 520)$ = 0.596 lb/ft^3

From equation (8-15):

$$\Delta P_{100} = 0.0000134 \times 0.00432 \times 100 \times (2292)^2/(0.596 \times 3.068^5)$$
$$= 0.19 \text{ psi} \qquad \text{ANSWER is b).}$$

Prob 44
From equation (8-4)

$$P_h = h\rho g/g_c = 6m \times \frac{1.1 \times 10^3 kg}{m^3} \times \frac{9.81m}{s^2} \times \frac{s^2.N}{kg.m}$$

$$= 64.746 \text{ kN}/\text{m}^2$$

Absolute pressure at the bottom of the column = 64.746 + 100 = 164.746 kN/m^2
ANSWER is b).

Prob 45
From equation (8-13) and from the conditions given by the problem:

$$\Delta P \propto Q^2$$

Therefore, $Q_2/Q_1 = (\Delta P_2/\Delta P_1)^{0.5} = (1.4)^{0.5} = 1.1832$
Increase in flow = 18.32%.   ANSWER is b).

Prob 46
From equation (10-11):
When D is constant $(Q_2/Q_1) = N_2/N_1$, and when N is constant for a geometrically similar fan, $Q_2/Q_1 = (D_2/D_1)^3$.
By the theorem of joint variation, when both N and D vary,

$$Q_2/Q_1 = (N_2/N_1)(D_2/D_1)^3$$

$$Q_2 = 7(30/20)(1.23/0.8)^3 = 38.16 \text{ m}^3/\text{s}, \qquad \text{ANSWER is c).}$$

Prob 47

By definition, key components are those whose separation is specified in the distillation. In this case, toluene is mostly taken as distillate product while ethylbenzene is removed from the column as bottoms product. Therefore ethylbenzene is the heavy key component.

Answer is b)

Prob 48

By doubling the tube side velocity, the tube side coefficient will be:

$$h_i = 125 \times 2^{0.8} = 217.64$$

$$U = \frac{1000 \times 217.64}{1000 + 217.64} = 178.73$$

ANSWER is a).

Prob 49

Component	Volume %	LFL %	UFL %	y_i	y_i/LFL_i	y_i/UFL_i
heptane	1	1.1	6.7	0.167*	0.152**	0.025
hexane	1	1.1	7.5	0.167	0.152	0.022
hydrogen	3	4	75	0.5	0.125	0.007
isopropyl alcohol	1	2	12	0.167	0.084	0.014
Total combustible	6			1.001	0.513	0.068
air	94					

*This is obtained as 1/6.   ** This is obtained as 0.167/1.1.

$LFL_{mixture} = 1/0.513 = 1.95$ %, and $UFL_{mixture} = 1/0.068 = 14.71\%$.

ANSWER is a).
Since the combustible volume % (6%) falls in the flammability range[1.95% and 14.71%], the mixture is flammable.

Prob 50

Time for sweep-through purging is given by:

$$t = \frac{V}{KQ} \ln\left(\frac{C_0}{C}\right) = \frac{5000}{0.2x3000} \ln\left(\frac{2.5x10^{-2}}{1x10^{-6}}\right) = 84.39 \ minutes$$

ANSWER is c).

Prob 51. The number of changes of container volumes = Qt/V = 3000x84.39/5000 = 50.634
ANSWER is d).

Prob 52

The equations necessary to compute the pH are:

$$pH = -\log_{10}\left(\gamma C_{H_3O^+}\right)$$

$$I = \frac{1}{2}\Sigma(C_i Z_i^2)$$

$$\log_{10}\gamma_{H_3O^+} = \frac{-0.5I^{0.5}}{1 + 3I^{0.5}}$$

Substituting the numerical values of concentrations and ionic charges:

$$I = \frac{1}{2}(0.01 \, x \, 1^2 + 0.01 \, x \, 1^2 + 0.09 \, x \, 1^2 + 0.09 \, x \, 1^2) = 0.1$$

$$\log_{10}\gamma_{H_3O^+} = \frac{-0.5I^{0.5}}{1 + 3I^{0.5}} = -0.0811388$$

$$\gamma_{H_3O^+} = 10^{-0.0811388} = 0.8296$$

$$pH = -\log_{10}(0.01 \, x \, 0.8296) = 2.08$$

ANSWER is d).

Prob 53

System : 100 kg of saturated liquid water in a closed vessel.
At saturation both pressure and temperature remain constant.
Energy balance reduces to $\Delta U = Q$
$P = 350$ kPa $\quad t = 138.3\ ^0C \quad$ from steam tables.
$x_i = 0.2$ and $x_g = 0.8$ by weight
$H_g = 2729.2$ kJ/kg $\quad$ and $\quad h_l = 581.6$ kJ/kg from steam tables
$\Delta U = Q = 0.8(\ 2729.2\ ) + 0.2(\ 581.6\ ) - 1.0(\ 581.6\ ) = 1718.1$ kJ/kg
Therefore, total heat added $= 100\ (\ 1718.1\ ) = 171810$ kJ
$\qquad$ Answer is (d).

Prob 54.

$$1 \ \text{ppm} \ = \frac{PM}{.08314T} = \frac{1.064 \times 17}{0.08314 \times 303} = 0.718 \frac{mg}{m^3}$$

$$180 \ \text{ppm} = 129.2 \ \text{mg} / \text{m}^3$$

$\qquad$ Answer is (b).

Prob 55

ANSWER is b)

Prob 56

ANSWER is a).

Prob 57

ANSWER is a).

Prob 58.

Synthetic division

$$
\begin{array}{rrrr|l}
1 & 4 & 6 & 4 & -1 \quad \text{trial root} \\
 & -1 & -3 & -3 & \\
\hline
1 & 3 & 3 & 1 & \text{a remainder of 1}
\end{array}
$$

$\qquad$ ANSWER is b).

Prob 59

$$x^{n+1} = x_n - P(x_n)/P'(x_n)$$

for $x_1 = -1$, $P'(x) = 3x^2 + 8x + 6 \mid_{x=-1} = 3 - 8 + 6 = 1$

$$P(x)\mid_{x=-1} = -1 + 4 - 6 + 4 = +1$$

$$\therefore x_{n+1} = (-1) - (1/1) = -2$$

ANSWER is d).

Prob 60

Using equation 13.9a

$$(x_{k+1}) = x_k - Tax_k + Tf_k \quad \text{for fixed increments.}$$

$$x_1 = x_0 - Ta_{x0} + Tf = 0 - 0 + 0.3 \times 1 = 0.3$$

ANSWER is c).

The most comprehensive review books available.

Engineering Press
The oldest and most respected source of FE/PE materials.

Engineer-in-Training License Review

Our Most Popular Book

Donald G. Newnan, Ph.D., P.E., Civil Engr., Editor

A Complete Textbook Prepared Specifically For The New Closed-Book Exam. All SI Units

Written By Ten Professors - Each An Expert In His Field
* 220 Example Problems With Detailed Solutions
* 518 Multiple-Choice Problems With Detailed Solutions
* 180-Problem Sample Exam - The 8-hour Exam With Detailed Step-by-Step Solutions
* Guaranteed! Pass the exam or your money back from the Publisher.

Here is a new book specifically written for the closed book EIT/FE exam. This is not an old book reworked to look new - this is the real thing! Ten professors combined their efforts so each chapter is written by an expert in the field. These include engineering textbook authors: Lawrence H. Van Vlack (Elements of Materials Science and Engineering), Charles E. Smith (Statics, Dynamics, More Dynamics) and Donald G. Newnan (Engineering Economic Analysis).

The Introduction describes the exam, its structure, exam-day strategies, exam scoring, and passing rate statistics. The exam topics covered in subsequent chapters are Mathematics, Statics, Dynamics, Mechanics of Materials, Fluid Mechanics, Thermodynamics, Electrical Circuits, Materials Engineering, Chemistry, Computers, Ethics and Engineering Economy. The topics carefully match the structure of the closed-book exam and the notation of the Fundamentals of Engineering Reference Handbook. Each topic is presented along with example problems to illustrate the technical presentation. Multiple-choice practice problems with detailed solutions are at the end of each chapter.

There is a complete eight-hour EIT/FE Sample Exam, with 120 multiple-choice morning problems and 60 multiple-choice general afternoon problems. Following the sample exam, complete step-by-step solutions are given for each of the 180 problems.
60% Text. 40% Problems & Solutions. 8-1/2 x 11, 2-Color,
ISBN 1-57645-014-7 5.0 lbs Yellow/Green **Hardbound** Cover
Order No. 147 $44.95 15th Ed 1998   784 Pgs

The fundamentals of Engineering/Engineer-In-Training (FE/EIT) exam has dramatically changed. For about the last fifty years the 8-hour exam has been a test of a dozen topics that are the core of every engineer's education. All that changed in the Fall of 1996. Now the exam has two distinct parts to it. The morning four hour portion is pretty much like the old exam with about the same dozen topics. The afternoon four-hour portion of the exam is new and different. In fact there are six different afternoon exams: civil, mechanical, electrical, industrial, chemical and a general exam. While the morning exam covers the first two years of an engineer's education, the afternoon exam covers about the last two years of the engineer's education in his specific branch of engineering. The morning exam has 120 single weight multiple-choice questions and the afternoon exam has 60 double weight questions in the discipline chosen by the applicant. We strongly recommend that an engineer take the discipline test in his own branch of engineering as these topics are those that the engineer will use most throughout one's career.

Fundamentals of Engineering Video Course

Thousands of engineers have viewed this video series.
Twelve - ½ hour programs prepared for Educational Television Network broadcast
A Video series that excels in graphics.
A golfer hits a ball with an initial velocity of 116 feet per second at an angle of 60 degrees, hoping to clear a grove of trees 10 yards away. The trees are 45 feet high and extend 320 feet in the direction of flight. Will the ball clear the trees? This is one of the many examples presented in the course.

Educational Television Broadcast Quality
Clear graphics - easy to see, hear, and follow
Only the more difficult aspects of each topic are presented, hence a concise review.
Presented as twelve ½-hour programs:

Introduction   Mathematics   Nucleonics   Chemistry   Statics   Dynamics   Mech
Mat'l   Fluid Mech   Thermo   Electrical   Mat'l Sci   Engr. Econ

The twelve programs are presented by eleven professors - each a specialist in his field. Students can use these videotapes as one of the easiest ways to review the topics. The 607 page Video Study Guide Book is also included. Use these videos to improve your test performance and your professional skills.

Thousands of engineers have viewed this video series. Now at a new lower price.
ISBN 0-910554-06-4   3 VHS Video Cassettes with Study guide. Boxed   5.5 lbs   1/CTN
Order No. 064   FE/EIT Video Course $99.00   + $9.50 shipping

Review Materials for the New Discipline Specific Afternoon FE Examination

EIT Civil Review

Donald G. Newnan, P.E.,Civil Engineer, Ph.D., Editor
Written by Six Civil Engineering Professors for the closed book afternoon FE/Civil Examination.
Each topic is briefly reviewed with example problems. Many end-of-chapter problems with complete step-by-step solutions and a complete afternoon Sample Exam with solutions are included.
Soil Mechanics and Foundations
Structural Analysis, frames, trusses
Hydraulics and hydro systems
Structural Design, Concrete, Steel
Environmental Engineering, Waste water
Solid waste treatment
Transportation facilities: Highways, Railways, Airports
Water purification and treatment
Computer and Numerical methods ISBN 1-57645-002-3

Order No. 139 $29.95 2nd Ed.1998   170 Pages

EIT Mechanical Review   Lloyd M. Polentz, P.E.,Mechanical Engineer

Written for the closed book afternoon FE/Mechanical Examination
Each topic is briefly reviewed with example problems. Many end-of-chapter problems with solutions and a complete afternoon Sample Exam with step-by-step solutions is included.
Mechanical Design, Dynamic Systems
Vibrations, Kinematics, Thermodynamics
Heat Transfer, Fluid Mechanics
Stress Analysis, Measurement and Instrumentation
Material Behavior and Processing
Computer and Numerical Methods
Refrigeration and HVAC
Fans, Pumps, and Compressors   ISBN 1-57645-004-X
Order No. 04X $29.95   1996   164 Pages   8-1/2 x 11

EIT Electrical Review   Lincoln D. Jones, P.E.

Written for the FE/EIT Afternoon Electrical Exam.
Each topic is briefly reviewed with many example problems with complete step-by-step solutions. Each chapter includes end-of-chapter problems with solutions. A complete Sample Exam with solutions is also provided.
The following topics are covered:
Digital Systems, Analog Electronic Circuits
Electromagnetic Theory and Applications, Network Analysis
Control Systems Theory and Analysis, Solid State Electronics and Devices,
Communications Theory, Signal Processing
Power Systems, Computer Hardware Engineering
Computer Software Engineering, Instrumentation
Computer and Numerical Methods   ISBN 1-57645-006-6
Order No. 066 $29.95 Revised Ed. 1997 176 Pages 8-1/2 x 11

EIT Industrial Review   Donovan Young, P.E., Ph.D., Industrial Engineer

Written for the FE/EIT Afternoon Industrial Exam.
Each topic is briefly reviewed with many example problems with complete step-by-step solutions. Each chapter includes end-of-chapter problems with solutions. A complete Sample Exam with solutions is also provided.
Production Planning & Scheduling, Engineering Economics
Engineering Statistics, Statistical Quality Control
Manufacturing Processes, Mathematical Optimization & Modeling
Simulation, Facility design and Location, Work Performance & Methods
Manufacturing Systems Design, Industrial ergonomics
Industrial Cost Analysis,Material Handling System Design
Total Quality Management, Computer Computations and Modeling
Queuing Theory and Modeling, Design of Industrial Experiments
Industrial Management, Information System Design
Productivity Measurement and Management   ISBN 1-57645-007-4
Order No. 074 $29.95 1996   154 Pages   8-1/2 x 11

EIT Chemical Review

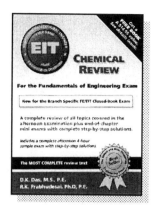

New

D.K Das,P.E.
R.K. Prabhudesai, Ph.D.,P.E.
Written for the FE/EIT Afternoon Chemical Exam. SI Units
Each topic is briefly reviewed with many example problems with complete step-by-step solutions. Each chapter includes end-of-chapter problems with solutions. A complete Sample Exam with solutions is also provided.
Material and Energy Balances, Chemical Thermodynamics
Mass Transfer, Chemical Reaction Engineering
Process Design and Economics Evaluation, Heat Transfer
Transport Phenomenon, Process Control
Process Equipment Design, Computer and Numerical Methods
Process Safety, Pollution Prevention   ISBN 1-57645-023-6
Order No. 236 $29.95 2nd Ed 1998 196 Pages 8-1/2 x 11

Additional Review Materials for the FE/EIT Examination

Fundamentals Of Engineering Study Guide
Lloyd T. Cheney, P.E.
Written by 10 professors. This is a 2-volume set: 288-page Study Guide plus a 144-page Solution Manual. Review of Math, Chem, Statics, Dynamics, Fluids, Strength of Materials, Thermo, Elect Circuits, Engr. Econ, Computer Sci and Sys plus 550 EIT problems with detailed solutions. 50% Text. 50% Problems and Solutions. Blue Soft Cover 8-1/2 x 11
2 volume set   Shrinkwrapped   3 lbs   12/CTN
ISBN 0-910554-70-9
Order No. 709   432 Pgs $19.50

Preparing For The **ENGINEER-IN-TRAINING EXAMINATION**

Irving J. Levinson, P.E.

640 E.I.T. Problems Each With A Detailed Step-By-Step Solution

Preparing For The Engineer-in-Training Exam
Irving Levinson, P.E.
640 Problems And Solutions For the National FE/EIT Exam
Mathematics, Physics, Chemistry, Statics, Dynamics, Mechanics of Materials, Fluid Mechanics, Thermo, Electrical Circuits, and Engineering Economics. 1992
100% Problems and Solutions.   6x9 Paperback   1 lb   22/CTN
Yellow/orange cover   **ISBN 0-910554-85-4**
Order No. 854   242 Pgs   3/ed $19.50

Being Successful As An Engineer
William H. Roadstrum, P.E.
Helps the young engineer make the transition from student to the practice of engineering. Describes engineering and how it functions within an organization. 6 x 9 Paperback   1 lb 36/CTN   Blue/white cover. **ISBN 0-910554-24-2**
Order No. 242 $14.95   1988   246 Pages

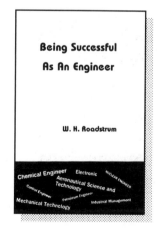

Being Successful As An Engineer

W. H. Roadstrum

Chemical Engineer   Electronic
Aeronautical Science and Technology
Control Engineer
Mechanical Technology   Petroleum Engineer   Industrial Management

College Reference Series

The Exam File Series are actual midterms and finals from teaching professors around the country. In each exam file you will find excellent review problems with the professor's own solution. Many of these are short and of the type used on the FE/EIT exams. Many students use these in their class studies and when preparing to take the FE exam.

Calculus I Exam File

29 Profs Reveal 301 Exam Problems
With Step-By-Step Solutions. Precalculus, Limits, Continuity, Derivative, and Integrals. **ISBN** 0-910554-61-7
Order No. 617   **250 Pages**   **$15.50**

Calculus II Exam File

27 Profs Reveal 356 Exam Problems With Step-By-Step Solutions
Transcendental Functions, Integration, Definite Integral, Conics, Polar Coordinates, Indeterminate Forms and Improper Integrals, Taylor Polynomials, and Series.
ISBN 0-910554-62-5
Order No. 625   **282 Pages**   **$16.50**

Calculus III Exam File

25 Profs Reveal 333 Exam Problems With Step-By-Step Solutions
Vectors in the Plane and Parametric Equations, Vectors in Space, Functions of Several Variables and Partial Derivatives, and Multiple Integrals.
ISBN 0-910554-63-3
Order No. 633   **282 Pages**   **$16.50**

Circuit Analysis Exam File

27 Profs Reveal 291 Exam Problems With Step-By-Step Solutions
Circuits, Networks, Transient Analysis, Complex Frequency, 2-Port Networks, State Variables, Fourier, and Laplace.
ISBN 0-910554-53-6
Order No. 536   **314 Pages**   **$16.50**

College Algebra Exam File

28 Profs Reveal 509 Exam Problems With Step-By-Step Solutions
Sets, Basic Algebra, Equations & Inequalities, Linear Functions, Polynomial Functions, Linear Systems and Matrices, Analytic Geometry, and Polynomial Eqns. **ISBN** 0-910554-77-3
Order No. 773   **410 Pages**   **$17.50**

Differential Equations Exam File

27 Profs Reveal 452 Exam Problems With Step-By-Step Solutions.
First Order; Second Order, Solution by Power Series, Laplace Transform, and Applications of Ordinary Differential Eqns. **ISBN** 0-910554-70-6
Order No. 706   **506 Pages**   **$19.50**

Dynamics Exam File

19 Profs Reveal 333 Exam Problems With Step-By-Step Solutions
Basic Concepts, Kinematics of Particles and Rigid Bodies, Particle Motion Relative to Translating and Rotating Frames, Kinetics of Rigid Bodies, and Vibrations.
ISBN 0-910554-44-7
Order No. 447   **346 Pages**   **$16.50**

Engineering Economic Analysis Exam File

16 Profs Reveal 386 Exam Problems With Step-By-Step Solutions
A 32-page review, followed by 386 exam problems and solutions. 30 Interest Tables. Previously published as two separate books: *Engr. Economy Exam File* and *Engr. Econ Review*. **ISBN** 0-910554-83-8 V2
Order No. 838V2   **296 Pages**   **$16.50**

Fluid Mechanics Exam File
12 Profs Reveal 203 Exam Problems With Step-By-Step Solutions
Fluid Statics, Motion, Momentum and Energy, Boundary Layer, Conduits, Compressible Flow, and Open Channel Flow.
ISBN 0-910554-48-X
Order No. 48X   218 Pages $14.50

Linear Algebra Exam File
27 Profs Reveal 437 Exam Problems With Step by Step Solutions.
Vectors, Linear Eqns, Matrices, Linear Transforms, Determinates, Eigenvectors, Eigenvalues, and Numerical Methods.
ISBN 0-910554-69-2
Order No. 692   442 Pages $18.50

Materials Science Exam File
21 Profs Reveal 311 Exam Problems With Step-By-Step Solutions
Structure and Behavior of Solids, and Engineering Materials.
ISBN 0-910554-57-9
Order No. 579   314 Pages $16.50

Mechanics of Materials Exam File
16 Profs Reveal 313 Exam Problems With Step-By-Step Solutions
Internal Forces, Stress, Stress-Strain, Axial Loading, Torsion, Bending, Deflection, Combined Stresses, Failure, and Columns.
ISBN 0-910554-46-3
Order No. 463   378 Pages $17.50

Organic Chemistry Exam File
27 Profs Reveal 560 Exam Problems With Step-By-Step Solutions.
ISBN 0-910554-66-8
Order No. 668   534 Pages $21.50

Physics I Exam File: Mechanics
33 Profs Reveal 328 Exam Problems With Step-By-Step Solutions
Vectors, Velocity and Acceleration, Equilibrium, Plane Motion, Work & Energy, Impulse & Momentum, Rotational Dynamics, and Harmonic Motion. **ISBN** 0-910554-54-4
Order No. 544   346 Pages $16.50

Physics III Exam File: Electricity And Magnetism
33 Profs Reveal 348   Exam Problems With Step-By-Step Solutions
Coloumb's Law, The Electric Field, The Electric Potential, Capacitance, Charges in Motion, DC circuits and Instruments, Magnetic fields, Magnetic materials, Induced EMF's and Inductance, AC circuits, Electromagnetic Waves and Radiation.
ISBN 0-910554-56-0
Order No. 560   346 Pages $16.50

Probability And Statistics Exam File
16 Profs Reveal 371 Exam Problems With Step-By-Step Solutions
Statistics, Probability, Discrete & Continuous Distributions, Variances, and More. **ISBN** 0-910554-45-5
Order No. 455   346 Pages $16.50

Statics Exam File
17 Profs Reveal 335 Exam Problems With Step-By-Step Solutions
Force Systems, Equilibrium, Structures, Distributed Forces, Beams, Friction, Virtual Work, and Other Problems. **ISBN** 0-910554-47-1
Order No. 471   346 Pages $16.50

Thermodynamics Exam File
12 Profs Reveal 238 Exam Problems With Step-By-Step Solutions
Covers all the topics of a typical first course in thermodynamics.
ISBN 0-910554-49-8
Order No. 498   250 Pages $15.50

Chemical Engineering License Review
D.K. Das P.E.M.S. Chemical Engineer

R.K. Prabhudesai, P.E. Ph.D. Ch.E. ☆☆☆☆
Prepared specifically for the PE Exam used in all 50 states.   146 Example problems with detailed step-by-step solutions.
BIG BOOK Format 8-1/2 x 11
Covers ALL topics on the Exam.
Easy-to-use Tables, Charts, and Formulas
Ideal Desk Companion to Perry's *Chemical Engineer's Handbook*
Complete References and Index.17 Chapters:
Units, Fluid Dynamics, Heat Transfer, Evaporation, Distillation, Absorption, Leaching,Liq-Liq Extraction, Psychrometry and Humidification, Drying, Filtration, Thermodynamics,Chemical Kinetics, Process Control, Engineering Economy, Plant Safety, Biochemical Engineering.

The introductory chapter review the test specifications and the author's recommendation on the best strategy for passing the exam. The first chapter review English and SI units and conversions. A complete conversion table is given. The next chapters review fundamentals of fluid mechanics, hydraulics and typical pump and piping problems. Chapter 3 covers heat transfer, conduction, transfer coefficients and heat transfer equipment. Chapter 4 covers evaporation principles, calculations and example problems. Distillation is thoroughly covered in chapter 5. The following chapters cover absorption, leaching, liquid-liquid extraction, and the rest of the exam topics. Each of the topics is reviewed followed by examples of examination type problems. This is the ideal study guide. This book brings all elements of professional problem solving together in one **BIG BOOK**. Ideal desk reference. Answers hundreds of the most frequently asked questions.
The first really practical, no-nonsense review for the tough PE exam. Full Step-by-Step solutions included.
ISBN 1-57645-000-7
Order No. 007   1996   2nd Edition   565 Pgs $69.50

Chemical Engineering License Problems and Solutions
D.K. Das P.E.M.S. Chemical Engineer & R.K. Prabhudesai, P.E. Ph.D. Ch.E.
    ☆☆☆☆
Prepared specifically for the PE Exam used in all 50 states.   **New**
188 New PE problems with detailed step-by-step solutions.
BIG BOOK Format 8-1/2 x 11Covers ALL topics on the Exam.
Easy-to-use Tables, Charts, and Formulas. Ideal Desk Companion to DAS's
Chemical Engineer License Review Complete References and Index.16 Chapters and a short PE Sample Exam.
Material and Energy Balances, Fluid Dynamics, Heat Transfer, Evaporation, Distillation, Absorption, Leaching,Liq-Liq Extraction, Psychrometry and Humidification, Drying, Filtration, Thermodynamics,Chemical Kinetics, Process Control, Mass Transfer, and Plant Safety.This is the ideal study guide. This book brings all elements of professional problem solving together in one **BIG BOOK**. Ideal desk reference. Answers hundreds of the most frequently asked questions. The first really practical, no-nonsense Problems and Solution book for the tough PE exam. Full Step-by-Step solutions included. ISBN 1-57645-011-2
Order No. 112 $29.50   1997   192 Pages   8-1/2 x 11 Hardbound

NCEES Sample ProblemsAnd Solutions: Chemical Engineering

The Official Publication Of The People Who Write The Exam

State Licensing Requirements; Description of the Exam; Exam Development; Scoring Procedures; Exam Procedures and Instructions; Exam Specifications; Twelve Sample Problems in Chemical Engineering and Solutions; Example Exam Materials.

Order No. 234   172 Pgs 1994 $30.00

For the latest information...
Check out our Web site at
Http://www.engrpress.com

How to Order:

Order by phone   (800)800-1651
Mon-Fri 9am-5pm EST

Order by Mail:

Engineering Press
P.O. Box 200129
Austin, TX 78720-0129

Include $4.75 for shipping within Continental US. Else call us.

Order by FAX. (800)700-1651
Include your name, address, credit card information.

Or contact your college bookstore.

About the Authors

Dilip K. Das, B.Sc(Honors),B.ChE(Honors), M.S.ChE, P.E is surrently a Pressure Safety Specialist of Bayer Corporation at Kansas City, MO, where he applies DIERS technology in emergency relief system design. Before joining Bayer, he was principal process engineer for Ciba Corporation at St. Gabriel, LA where he serves as an interdivisional consultant with responsibilities including process design, simulation, technical support of production, cost control and project management. He has worked for Rhône-Poulenc, Inc. as project manager of agro-chemical projects. He has also worked for Stauffer Chemical Company as senior process engineer and for C. F. Braun Company as senior engineer. He received an M.S. in Chemical Engineering from the University of Washington and a B.Ch.E. (honors) in chemical engineering from Jadavpur University, Calcutta, where he received gold medals for excellence. Mr. Das has several publications and a US patent to his credit.

Rajaram K. Prabhudesai, Ph.D., P.E., is currently a consulting chemical engineer. He worked as process supervisor for Badger Engineers where his responsibilities were process and plant design. Previously he was senior process engineer with Stauffer Chemical Company, where he was responsible for process design, systems engineering, process economics, and plant start-up. Also he worked for AMF as principal research engineer and for Coca Cola Company as principal chemical engineer. He has written numerous papers for professional journals and contributed to Perry's *Chemical Engineering Handbook* and Shweitzer's *Handbook of Separation Techniques for Chemical Engineer* (both McGraw-Hill) and co-authored *Chemical Engineering License Review* (Engineering Press). Dr. Prabhudesai received his M.S. from the University of Bombay and his Ph.D. in chemical engineering from the University of Oklahoma.